Leaving Us to Wonder

SUNY series in Philosophy and Biology
David Edward Shaner, Editor

Leaving Us to Wonder

An Essay on the Questions Science Can't Ask

Linda Wiener

and

Ramsey Eric Ramsey

STATE UNIVERSITY OF NEW YORK PRESS

Published by
State University of New York Press, Albany

Cover photo courtesy of the National Gallery Picture Library

For information, contact State University of New York Press, Albany, NY
www.sunypress.edu

Production by Diane Ganeles
Marketing by Anne M. Valentine

Library of Congress Cataloging-in-Publication Data

Wiener, Linda, 1957–
    Leaving us to wonder : an essay on the questions science can't
ask / Linda Wiener and Ramsey Eric Ramsey.
        p. cm. — (SUNY series in philosophy and biology)
    Includes bibliographical references (p.) and index.
    ISBN 0-7914-6313-3 (alk. paper) — ISBN 0-7914-6314-1 (pbk. :
alk. paper)
    1. Philosophy and science.   I. Ramsey, Ramsey Eric, 1960–
II. Title.   III. Series.

B67.W54 2005
501—dc22   2004048161

10 9 8 7 6 5 4 3 2 1

Contents

Preface

This all began before we met, of course. One of us trained in biology and the other in philosophy. It is likely if we were both to have followed the seductions of our initial studies and trainings, then we would in all likelihood have been enemies in today's so-called science wars. As it stands we both set out on paths diverging from those original beginnings and today we share a similar set of concerns that make us allies. Granted, in this case the allying is made possible in large part by the biologist becoming a philoso-pher as well. Be that as it may, this is not a joining of forces in favor of one camp or the other. Since our first meeting under the auspices of the Ometeca Institute—which seeks to bring together those persons practicing in the sciences and the humanities—we have attempted to create a philosophical way of life together. That is to say, by cultivating friendship through meditating on ques-tions of living well in an epoch whose major manner of being is techno-scientific, together we have fostered an ongoing dialogue. We are not so bold as to think that our friendship nor this collaboration will bring anything like a peace settlement to the science wars, yet we are hopeful that the practice of dialogue rather than mere polemics might be a real, even if small, contri-bution to something worth pursuing.

Along with Ometeca's founding members and its director Dr. Rafael Català, others interested in such collaborations have made opportunities for us to present aspects of our work to other interested persons. A presentation at Harvard University's Depart-ment of Botany was made possible by Peter Stephens. The event was attended by a number of Harvard biologists and their stu-dents. According to their reports, both groups found the work novel and engaging. Perhaps most important, the younger scien-tists in attendance were pleased that many of their persisting questions (but ones left unaddressed by their scientific training) found expression in our work. Many of these students believed our approaches to the questions allowed them to pursue rather than ignore their fundamental concerns.

Yael Lubin of Ben Gurion University, Israel arranged for an important discussion of the work. As with the other scientific venues, this event was attended by members of the scientific community and others working at Ben Gurion's field station in Sde Boqer. Again many found the debate as we put it forward a significant and cogent challenge to mainstream thinking about science, determinism, and public life.

In a similar vein, few of the scientists working in the Genetic Research Group at Los Alamos National Laboratories came with any sympathies for our position. Nonetheless, in the lengthy discussion that followed many found themselves dealing with familiar topics in new ways because of the position we advanced. It is safe to say no minds were changed on this day, but habitual ways of understanding the issues where indeed challenged. Moreover, we learned much from having to address their many concerns.

Philosophers Martin Beck-Matustik and Patricia Huntington made possible a public reading of a version of our final chapter at the Center For Theoretical Study, Prague, Czech Republic. The Center's working group of scientists and philosophers offered us many fine insights concerning issues fundamental to the work. We remain grateful for this wonderful opportunity and for the generous insights offered to us by all those in attendance.

We thank William Homestead whose reading, late in the process, corrected mistakes and forced clarifications of parts of the position that otherwise would have remained obscure. His fine suggestions and questions made for a better text despite what faults might remain. Cynthia Gaffney also gave a valuable reading late in the process. In addition, Robert W. Jones offered his expertise concerning the writing and presentation of this project in its many stages and with respect to its various audiences. At various stages of the work a number of undergraduates at Arizona State University, West Campus assisted in the preparation of the manuscript: Jennifer Anderson, Barbara Giles, Carol Marshall, Sohinee Roy, and Daniel Smith. We also recognize the administrative support of Dawn Poh. We are thankful for the good work of Liz Marini who prepared the bibliography and the index. In securing the cover art, Charles Lee Mudd's expertise was invaluable.

The project has benefited immensely from the flawless guidance of Jane Bunker, our editor at SUNY Press. Her belief in this project breathed new life into it when doubts about it hung heavy in the air. In addition to Ms. Bunker's support, the production of the text was overseen by Diane Ganeles with patience and care.

RER would like to acknowledge his gratitude to and for the friendship of Heidi Bostic, Richard Dienst, and Stephen Pluhacek who make a cell of resistance a living reality. Furthermore, he knows that this work would not have been possible without Diane Gruber whose being-there-with is a beautiful sharing of everything that matters. The vastness of her contributions cannot be calculated.

Chapter One

A Place to Begin

The person asking questions is merely exercising the right that has been given him (in serious dialogue): to remain unconvinced, to perceive a contradiction, to require more information, to emphasize different postulates, to point out faulty reasoning, and so on.

—Michel Foucault, "Polemics, Politics, and Problematizations"

Curiosity and Wonder

Ultimately, are not wonder and curiosity poles apart?

—Henry Bugbee

Always, no matter how much or how often we satisfy our never-ending curiosity with facts, something profound remains untouched. That which remains—something far apart from curiosity—is the experience of wonder. We might call our experience of this, and thus harken back to the ancient Greeks, suffering wonder. We are persuaded when it comes to asking how we ought to make a way of life together no number of facts can provide a risk-free blueprint for living well. This is not to say, of course, that we should be done with curiosity and facts or with the practices and the institutions of science that produce the vast majority of our factual knowledge. To the contrary, we may never have enough facts for our curiosity, and yet curiosity and facts are never and never will be enough to answer all our pressing questions. The practice called philosophy understood in its broadest and least professional sense can provide, albeit without any guarantees, the risky and responsibility-laden understandings that we turn toward these perennial questions surrounding our attempts at living well.

This way of wonder is, thankfully, open to all because each of us, as we move through the world, already has and expresses a philosophy. However, to have a philosophy does not necessarily mean one has chosen it. Many of us find ourselves with philosophies, and they remain in a strange sense both ours and unknown to us. This situation can be overcome, we believe, by engaging in various and varied philosophical exercises. Briefly stated, the practice of philosophy advocated herein sets itself on the path of continual attempts to reflect on the form, content, and consequences of the philosophies we choose to hold. We shall pay special attention to those positions that in this techno-scientific age, with all its supporting structures from education to popular culture, are disseminated everywhere and always. Engaging in philosophy will be seen as a radically important endeavor as we realize that when they remain uninterrogated, the philosophies we currently hold can have a cruel hold on us.

The story is a long and complicated one so much so that there are many possible versions of it. It is a story abundant in detail as well as one filled with numerous contestations and conflicts. In light of this, we shall tell but one version of this rich narrative. Because the practices of science are the foremost producers of facts, we tell our version in a manner that focuses on a certain type of Western scientific thinking. All the while, of course, we have our eyes on a path that leads to an understanding of the essential place philosophy broadly thought might be said to have still in a techno-scientific age.

A Place to Begin

By the mid-eighteenth century, The Royal Society of England, that great institution of modern science, was already firmly established. It was an exciting time for European science. Physicists and chemists were discovering the laws that govern the behavior of gasses and the composition of bodies. Rigorous experimentation coupled with elegant mathematical equations were changing our notions about nature and giving us an unprecedented control over its secrets.

It was during this time that the British artist Wright of Derby painted *The Experiment on a Bird in an Air Pump*, illustrated here. The traveling representative of science, half in light, half in shadow, has put a white-feathered bird into the glass globe of the air pump. Upon the evacuation of the air from the globe, the bird will die, thus illustrating the then-still novel phenomenon of the vacuum. The child and her family are strongly illuminated; she has a stricken look, her mother cannot bear to watch, and her father patiently explains the theory of the vacuum. The other onlookers are curious, thoughtful, yet the two lovers turn their interests and thoughts toward one another.

We can see the four most prominent figures, those focused in the center of the painting, as symbolic of the issues and concerns we wish to address in what follows. The experimenter's piercing and seemingly single-minded look dominates the scene, indeed is directed at us. His gaze seems symbolic of an unswerving conviction in the power of the newly developing theory; his is the hand poised to let the bird either live or die based on which choice he makes, how he chooses to let the inevitable laws of nature play themselves out. This much we know, without oxygen the bird will perish and access to this oxygen is dependent on his hand, but the laws of the physics of the vacuum are inevitable. His look suggests a fascination with the power that knowledge of these laws gives him. The bird's death or even near asphyxiation proves the power of the physical fact to the as yet uninitiated public.

The experimenter alone is not enough. He needs disciples in the public realm to spread the new doctrine. This public face is exemplified by the husband. He consoles the woman who cannot bring herself to watch the bird's demise at the hand of the experimenter, nor watch its suffering. Perhaps she believes that it does not require the risk of the bird's breath to prove the science of the vacuum. In any case, the consoler's arm around her shoulder tells her "everything will be all right."

These two men at the center of the painting, the collaborators, are akin to the two aspects of science that we shall explore. The first is the undeniable beauty of pure theory, prediction, control, and reduction to the inevitability of the laws of nature. The latter, the consoler, represents the repeated claim that somehow it will turn out fine and that we (though perhaps not the bird) must in the end succumb to the inevitability of the theory and the law. We cannot help but wonder if the reassurance from the latter conflicts with the convictions of the former? The consoler does not have the scientist's power, but the scientist needs the public to believe in and support his endeavors. For different reasons they both want to secure the conviction of the woman whose eyes cannot meet those of the bird as it becomes a scientific experimental object. Or perhaps she is consoled; in the end she may come to believe.

Regardless of the woman's ultimate position on this, there is yet another face and another look. The young girl, close enough to the consoler to hear his comforting words, does not turn his way. She looks back to the hand that is raised above the cage that traps the bird. She indeed feels for the bird and the viewer's thoughts come back around to what this might mean for her and for us.

Wright is clearly concerned with the power this new science has to elevate the value of knowledge over the value of life, over the humane value of recognizing a little girl's concern for the bird. Wright shows us that the objective world of science cannot be kept separate from the world of our day-to-day lives. Historically, science and the technology that comes with it have had the power to reorient our priorities and introduce new values into our lives. In the painting, though the pain of the little girl and her mother are recognized, no one stops the experiment. It is as if truth as a value trumps what concern one could have for the experience of those onlookers who have yet to be won over.

Mistaking the Abstract for the Concrete

There is a widespread belief, held to be common sense, that one great virtue of science and mechanistic analysis is that it gets to what is real or to reality itself and that it provides the only concrete descriptions and explanations we can have. Following from this belief is an assertion that those privileging literary, philosophical, or religious manners of understanding are, at best, engaged in abstract imaginings and, at worst, ideological propaganda. Contrary to being abstract, ours is an analysis that suggests our lived experience is the most concrete thing with which human thinking could begin.[1] Consequently, literature, philosophy, or religion that originate from our engagement with the world can be seen as more relevant to our concrete lived experience than the "reality" that science provides.

The American pragmatist John Dewey shares the following example that gets to the heart of this issue. In *Experience and Nature* Dewey writes: "To pass over in [the practice of] science the human meanings of the consequences of natural interactions is legitimate; indeed it is indispensable."[2] On this account, we begin not with the facts of science but with our lived experience and then abstract ourselves from it *so that we can be scientific.* But this is not getting closer to reality; indeed, it moves in the other direction. Science is one of the most abstract social practices because it can, and *qua* scientific practice it is obligated to, bracket the lived world of experience so as to have room to do its work.

To demonstrate this point Dewey argues that the formula H_2O is an abstraction. This formula can say what it does about water, not because that is what water "really" is, but because water is already given to us in our lives and is first meaningful to us as we come to know it *within* the concrete dealings of our lived experience. Knowing water scientifically is one possible experience among others, nonetheless it remains one that comes to us, and must necessarily come to us, as an abstraction. Water is concretely meaningful because people use it in some of the most important ceremonies of their lives; it also has its concrete meaning in its literary and symbolic uses in art that have nothing to do with it being understood as H_2O. These lived meanings are the elemental meanings of water as it is understood and experienced in the fullness of a larger web of social relations.

Dewey links this understanding of truth and abstraction with our experiential concern for meaning when he writes:

> Truth is a collection of truths; and the constituent truths are in keeping with the best available methods of inquiry and testing as to matters-of-fact; methods, which are, when collected under a single name, science. . . . But the realm of meanings is wider than that of true-and-false meanings; it is more urgent and more fertile. When the claim of meanings to truth enters in, then truth is indeed preeminent. But this fact is often confused with the idea that truth has claim to enter everywhere; that it has monopolistic jurisdiction. Poetic meanings, moral meanings, and a large part of the goods of life are matters of richness and freedom of

meanings, rather than truth; a large part of our
life is carried on in realms of meaning to which
truth and falsity as such are irrelevant.[3]

Phenomena are rich in their possible meanings and one
perspective used to understand them is seldom, if ever,
adequate, especially when that method is the abstraction of
science. As Dewey makes clear the claims to Truth do not
enter in everywhere, and for us one place where they ought
not enter in as preeminent is the social space called the
"public sphere" where we deliberate and reflect on questions of the common good as they relate to our shared lived
experiences of the social world.

There are reasons for our concern that claims to the
Truth threaten the public sphere. In an article in the highly
respected popular science monthly *Scientific American*, we
read the following promotion of evolutionary psychology
that suggests our concern is justified:

> Darwinian science inevitably will, and should, have
> legal, political and moral consequences; some of
> the most pressing issues . . . —abortion, birth con-
> trol, sexual discrimination, homosexuality—are "on
> Darwin's beat." . . . at least one Supreme Court jus-
> tice and several high-ranking Pentagon officials . . .
> have taken an interest in evolutionary psychology
> and are considering applying it in their realms.
> Ready or not, here comes the Darwinian society.[4]

There are similar assertions for the importance of "Darwinian" thinking in medicine, law, and politics as well as many
other aspects of our lives. For example, Daniel Dennett
argues, using Darwinian principles, that we should relegate
religions to the status of mere relics.[5] He believes that science is our only hope for the future and if we do not "quar-
antine" religion, then we shall seriously threaten our future.
Despite Dennett's best intentions, it is difficult not to see

this use of science, rationality, and truth as having danger-ously authoritarian leanings.

In the light of these examples, we think it not hysteri-cal, alarmist, or irrational to dissent not only on intellectual grounds, but also in the belief that these ideas are being disseminated, adopted, and advocated far too quickly and enthusiastically in the public sphere. Surely it is pressing social issues for which the most careful meditative thought is called. In the absence of the time, access to necessary tools of thought, and even the inclination of most citizens to think these issues through, there is the danger that such authoritarian scientific views will be taken up and acted on by those who have political power. We can only hope that it will not be in retrospect that we have to acknowledge the danger and power that such widely disseminated ideas may wield in our communities.

We can find in Wright's painting a perspective from which to raise questions concerning our experiential lives and the life described for us by science. Can moral or ethi-cal prescriptions follow necessarily from the facts given to us by scientific investigation? A quick look at three well-publicized scientific works that attempt to use scientific meth-ods as the means to think and also to ground claims about the ultimate workings of the social world provides us with evidence of the distance between scientific abstraction and the concrete meanings of our lived experience. In these three cases E. O. Wilson, an expert on ants, pronounces the whole world to be selfish; James Lovelock, from his studies of bacteria, sees the great whole of Gaia as a self-regulating system and urges the truth of cooperation; while Humberto Maturana and Francisco Varela, from a physiological per-spective, urge that only love can preserve our "autopoiesis."[6]

Concerning these three conflicting biological perspec-tives there are two important issues to be raised here. The first is that at the level of science each of these projects produces sound scientific analysis. This is to say, each has sets of data that were produced within the dictates of the scientific method. The results are repeatable, verifiable, and lead to exciting questions to be undertaken as a continuation of the research project each has initiated.

Problems show themselves here because of the widespread assumption that somehow ethics must have a scientific foundation. These three scientific projects do not share a unified set of moral prescriptions. How is it that three research projects that all hold to the highest standards of producing scientific truth are then at odds with one another when we compare their respective moral claims? Why have three bodies of truth production not produced one set of moral prescriptions as one's faith in science might well expect? How do we decide now which of these pronouncements is correct? Surely not on the basis of the scientific facts. What is demonstrated here is that, on their own shared commitments to science and its ability to inform us about ethics, these three positions face troubling divergences in opinion when they present their competing and incompatible moral claims as scientifically based.

A second issue that emerges from consideration of these three projects together is that each, when it turns to philosophy, does so without taking philosophy seriously.[7] These authors assert versions of philosophical positions in manners that suggest philosophy is immediately accessible to anyone, whether or not the reader has done any of the philosophical work necessary to produce a rigorous and sophisticated position. Books with page after page of wonderful science writing and thinking end with a few pages of watered down philosophy and social theory that ignores—in ways their scientific work would never—long traditions, other relevant insights, and important debates surrounding the philosophical principle they have chosen to embrace. In these three cases selfishness, cooperation, and love all appear on the scene as if they had no intellectual history and emerged as *necessary conclusions* from their scientific work. However, it is clear from the passionate rhetoric and prolific popular publications of scientists that these philosophical views are a vital, though perhaps not a well-examined, part of their intellectual lives.

We criticize scientists for what is perhaps best characterized as sweeping and unexamined philosophical claims
that are intellectually appealing because of their simplicity
and because they privilege science and its methods as the
only road toward any sort of reliable truth. From this perspective, reason and scientific method claim to overcome
previously relied on superstition and irrationality of all sorts.
We wonder if these claims are justified.

Today's Popular Science

It is absolutely safe to say that, if you meet somebody
who claims not to believe in evolution, that person
is ignorant, stupid, or insane (or wicked, but I'd
rather not consider that).

—Richard Dawkins, *The Blind Watchmaker*

There is indeed a force devoted to enticing us into
various pleasures that are (or once were) in our genetic interests. . . . You could call that force the "ghost
of natural selection." More concretely, you could call
it "our genes" (some of our genes at least). If it will
help to actually use the word *evil* there's no reason
not to.

—Robert Wright, *The Moral Animal*

Inability to think is not stupidity; it can be found in
highly intelligent people, and wickedness is hardly
its cause, if only because thoughtlessness as well as
stupidity are much more frequent phenomena than
wickedness.

—Hannah Arendt,
"Thinking and Moral Considerations"

Considering the first two of the above passages from Dawkins and Wright, we confront some modern examples in which the values and views of scientists make their way into the lives of nonscientists. Both these quotations are from books about evolutionary theory written for the lay public. It might strike a reader as strange that even small stretches of a long deoxyribonucleic acid molecule could be termed evil, and furthermore, that persons who do not believe in the absolute effective power of large organic molecules could be called wicked. To put this question in a larger context: Why is it that such emotionally laden language is being used by scientists and promoters of science to convince us of the truth of their theories?

Far from being idle concerns or simple curiosities, we believe these questions illuminate some of the most important concerns for all of us in this age of science. Ethical language appears in the works of scientists that are written for other similarly trained scientists and is propagated in the work of journalists or scientists themselves writing for the general public.

The mainstream popular press including such monthly and weekly periodicals as *Newsweek*, *The New Yorker*, and *The Atlantic Monthly*, as well as the daily press, has published a barrage of articles on the subject of Darwinian science. Seldom does a week pass by without a headline article on some aspect of this subject. We are told that we and all living creatures are mere vehicles for the survival of selfish genes: "DNA neither cares nor knows. DNA just is. And we dance to its music." We are assured that once we realize that DNA is all humans are "for," we can use this insight to help us "reach goals, choose goals, distinguish practical and impractical goals and decide which goals are worthy of our pursuit."[8] There is something rather cavalier in all of these assurances that may be far from reassuring to those who do not share these authors' presuppositions and foundational claims about the ultimate cause of human behavior. Their presuppositions seem to deny human agency and freedom, two qualities that would seem to be necessary if one is going to be responsible for making ethical and moral claims.

Those of us who do not share these authors' presuppositions may be excused for wondering why we should accept these "findings" or strive to live according to these "insights." Richard Dawkins scolds those of us who are tempted to ask such audacious "why" questions. In his words: "Now only the scientifically illiterate" ask "why" questions "where living creatures [are] concerned." He continues: "But 'only' conceals the unpalatable truth that we are still talking about an absolute majority of the world's population."[9] We are particularly interested in why scientific writing for the lay public finds itself so closely intertwined with ethical and moral language. Is this simply a common rhetorical device, or is it part and parcel of the scientific arguments themselves?

As a first approach to these questions, we suggest that this close connection comes about because the ethical and material realms seem to be both in intimate relation with one another and, simultaneously, at odds. If we were to put our immediate concerns into the form of questions, we would find that the list includes at least the following:

1. What are the relations between scientific knowledge and ethical conduct?
2. Is scientific knowledge the same as truth?
3. If it is, what is the relation of truth to ethics?
4. How do we reconcile the knowledge gained from a scientific assessment of the world with the broader concerns that confront us in our day-to-day lives where we act as if we are in some sense free?

Granted these questions are not new in the history of Western thought. Yet, the influence of science in our time ought to bring these familiar questions into a sharp contemporary focus. Indeed, almost all of our actions in the world imply answers to each of these questions even if they are seldom, if ever, explicitly raised. That is to say, our actions embody answers to these questions and as such we ought to know what answers we are giving by reflecting on these issues.

One example that brings many of these issues to the fore is that of Alan Sokal, a physicist at New York University who published the paper "Transgressing the Boundaries: Toward a Transformative Hermeneutics of Quantum Gravity" in the leftist journal *Social Text*.[10] The style of this essay imitates the worst excesses of postmodern writing while presenting claims that certain "findings" in the scientific field of quantum gravity support the postmodern view that truth is relative. The science "portrayed" in Sokal's essay was very often amusingly wrong, and yet it was presented in the typical style of scientists who seem to think that everyone worth communicating with should have a working knowledge of their particular discipline.

This paper, as Sokal admitted concurrently in the journal of academic life *Lingua Franca*, was a hoax meant to dramatize the vacuousness and ignorance of the postmodern critics of science. The responses to this hoax have been various—from a condemnation of Sokal's breech of professional ethics to wild self-congratulation by scientists—"Now," they exulted, "we have finally got them!"

Less noted by those responding to this so-called hoax was Sokal's commentary on "correct" science that took the form of extravagant and insincere compliments to scientists espousing unorthodox views, and extravagant and insincere vitriol directed against scientists who support the banner of science against any and all of its critics. Apparently, only scientists who support a particular view of science are really in the club. Missing from Sokal's and other scientists' criticisms of those who do not share their views of science is any self-critical effort to assess whether there are substantive issues that they themselves may not have adequately considered or which they do not understand.

The Sokal incident has led to much public discussion of the relations among science, philosophy, and social criticism. For some, this event marked the beginning of the discussion, as many who do not follow such debates first heard of the "science wars" and their escalation when the event was covered by large numbers of news and opinion weeklies and daily newspapers. Others, who follow such things closely and were sympathetic to Sokal and his ruse, see it as the end and the final word, believing (or at least hoping) that this event will have shamed those who question science into silence. Yet to our minds, neither of these views quite gets it right. The Sokal event and its aftermath adds no new insight and states no new positions. The science wars, though now brought to the bright light of public discourse, remain, when seen from the perspective we develop, in the same standoff positions as before.

This standoff is exemplified, we think, by the meeting that occurred between principles (the editors of *Social Text* and Sokal) after the hoax. *The New Yorker* reports that Sokal presented two versions of how Native Americans came to occupy a particular place in North America, one the current scientific account from archeological evidence and the other the mythical account given by a Native American tribe and then asked which was true? The audience, led by *Social Text* editor Andrew Ross, responded by questioning the question. In essence he asked: "Who wants to know?" and "Why do you ask?" According to the article, these questions in answer to his question puzzled Sokal—he was thinking, no doubt, that the answer was obvious and the questions themselves showed the unwillingness of members of his audience to see the truth so obviously before them. Sokal assumed that the context of the scientists is automatically assumed to be the universal context in which such questions could be asked and answered.

Sokal's most recent book *Fashionable Nonsense* contains much cogent criticism of the use of science in what he classes postmodern thought. Nonetheless, he holds to his above noted position. This is exemplified when he remarks: "It is perfectly legitimate to turn to intuition or literature in order to obtain some kind of nonscientific understanding of those aspects of human experience that cannot, at least at present, be tackled more rigorously."[11] Sokal indicates here

that science is the final arbiter of all questions and all literature can do, it seems, is to help us bide our present time
until science gets to its next set of rigorously won truths.

This view of Sokal's does not belong to him alone,
indeed it is shared by many others: *that science is the proper
and exclusive foundation for thinking about and answering every
question.* We follow others who call this "scientism," a world
view characterized by its authoritarian attitudes, its totalizing drive to encompass every question, and its disregard
and disdain for alternative views.[12] Thus, whether a piece of
research is either good or bad science is rightly judged from
the perspective of scientific rules and conventions. By
scientism we understand a way of thinking and the public
statements associated with it that extend the legitimacy of
scientific thinking as such to issues and contexts outside the
purview of science as a practice. A most troubling claim and
one central to scientism is made by nearly all the scientists
whose work we examine here, namely, that science has its
limits. Yet this proclamation does not then restrict, in fact it
sets up, the transgression of these limits that inevitably follows. Consequently, there is always one line of text that
proclaims the recognition that science has its limits, to which
all critics are referred, that is meant to act as some sort of
vaccination against critique. But recognition is not respect;
one can recognize limits and still transgress them as do
most of the scientific authors we cite. As John Lyne puts it
with respect to his critique of E. O. Wilson's *Sociobiology:*
"Genetic determinism was affirmed in image and story, even
while being denied in theory."[13] We shall not be silenced by
referral to one or two lines in books and articles that, as a
whole, already have no respect for their own caveats.

The current hotly contested debate between science and
its critics suggests that we ought to take these questions seriously even if we feel we have heard them before. Our overall
goal is to explore these questions by refocusing them, by
viewing them through some intellectual lenses not usually

associated with these debates. We take evolutionary biology, specifically its recent sociobiological and evolutionary psychological forms, to be an exemplar for our reflections. We will see that debates over evolutionary biology take us to the heart of many broader and more fundamental issues raised by our scientific age.

To achieve this refocusing, we need a backdrop against which to see the importance of our questions. We are persuaded that we can learn much from asking and attempting to answer questions that may turn out to have many answers or even remain perpetually unanswerable. We wish to give those not familiar with the various traditions of philosophical discourse access to new tools for thinking about these and other questions. These philosophic tools are not only valuable for the new insights they can give, but are necessary conceptual understandings if we hope to understand humanity's place in that world that science describes for us. So as not to be mistaken let us repeat, we believe scientific thinking plays a critical role in our pursuit of understanding and we offer reasons why other perspectives, while not negating or disregarding the scientific, also *must* play an essential role. Consequently, we are not focusing on the intra-scientific issues of logic, statistics, or experimental methods, we are instead willing to take them (for the most part) at face value. We rather want to step outside the scientific domain to raise questions that we believe cannot be asked from the inside.

The distinction we are drawing between philosophical thinking and scientific reasoning is central to our task. To make a place for these reflections, we turn here to the spirit of Socrates to emphasize the gulf between the type of argumentation characterized in the first two epigraphs at the start of this section and the type of philosophical reasoning we are highlighting. This allows us to invoke the spirit of the great Athenian philosopher whose style and motivation for questioning and investigating all aspects of our lives ought to be a living part of our intellectual heritage.

The Spirit of Socrates

In the Platonic dialogues, the subjects with which Socrates and his interlocutors deal range from the nature of the physical world to questions of knowledge, justice, truth, virtue, and love. While inquiring about these matters, Socrates spoke with slaves, craftsmen, teachers of rhetoric, powerful politicians, students, and anyone else willing to raise and explore questions. One aspect of Socrates' character we particularly want to stress is his willingness to trust in the capacity of his interlocutors. To this end he claimed to be a midwife—helping his friends to bring the truth to fruition within themselves and helping them to give birth to their own ideas rather than insisting they bow to his authority. Indeed, logical reasoning plays as large a role in the dialogues as it does in modern science, philosophy of science, and analytic philosophy, but missing from these latter fields is the constant questioning of all presuppositions, especially the inquiry about highest human goals and the way to lead the best life. It is these latter types of questioning that, in our view, quintessentially define the task of philosophical thinking.

Here Socrates' passion for certain types of questions is instructive. From the question "Is it better to do a wrong or suffer a wrong?" in the dialogue *Gorgias* to the question motivating the *Republic* "What is justice?" Socrates asks questions that go to the heart of our lived experiences and our relations with others. He does not ask out of idle curiosity but because he believes the answers to these types of questions can lead us toward living well and freely rather than living as slaves to ignorance. Living well includes not just searching for answers to these questions, but taking a certain pleasure in the asking itself and pleasure in the ensuing dialogue and conversations with other human beings. Part of living well is having the courage to engage in just these types of questioning dialogues.

In contrast to many of the scientific thinkers with whose work we deal in this essay, Socrates never gets angry with those who disagree with him, never resorts to ad hominem arguments, never doubts people's ability to understand, and never refuses to speak with anyone interested in dialogue. It is obvious that he is not threatened by those who disagree with his positions because he is interested in the truth, goodness, and fruitfulness of the idea, and does not judge it by the one who holds it. Perhaps most important for our current purposes, Socrates always believes himself free enough to ask "Why?"

The sense of philosophy that is at work in our essay is indebted, even if it cannot live up fully, to the Socratic spirit we have just described. Thus, when we turn to other philosophers and thinkers who in some sense share this spirit, it is to invite examination of sometimes familiar questions from new points of departure.

No doubt the means of delivering the newest scientific discoveries have changed greatly since the time of Wright's painting with which we began. Today, for those not involved in scientific practice, such demonstrations occur in the popular press propagated by scientists themselves and those reporters eager to take such news to the general public. Certain similarities remain between the two times if we but imagine ourselves gathered around the latest book or article that seeks to promote the newest scientific truth. In this light we may well wonder: who or what shall comfort us as we observe each successive demonstration?

In the next chapter, we first offer a brief historical overview, examining the various ways in which important Western thinkers have conceived of the relationship between the material and ethical realms, between right knowledge and right action. Informed by this discussion, in the third chapter, we examine the evolutionary and moral thought of Darwin and Nietzsche, two of the most influential thinkers of the nineteenth century and founders, respectively, of major theoretical schools in contemporary biology and philosophy. Lastly, the final three chapters draw on contemporary philosophical positions that take both science and our lived experience of the world seriously. We hope our study, which proceeds with equal respect for science and philosophy, will act as an invitation to dialogue. We are attempting to create a public space in which informed discourse might take place where what comes to matter to us are not only questions of truth but more importantly questions of living well.

Chapter Two

The Persistence of a Question

What interests everyone, or rather what *should* interest everyone, is nothing other than wisdom. The normal, natural, everyday state of human beings ought to be wisdom, but they cannot reach it; this was another of the fundamental ideas of ancient philosophy. It amounts to saying that what interests each person is not only the Kantian-critical question "What can I know?" but especially questions like "What should I do?" "What may I hope?" "What is humankind?"—which are the fundamental questions of philosophy.

—Pierre Hadot, *What Is Ancient Philosophy?*

Some Background Questions

Our interest is in the workings of the material world and its relation to the intellectual, moral, and in the broadest sense of the word, religious life of humankind. The contemporary discussion that we are entering is yet another round in a debate that goes back in our culture at least to the Ancient

Greeks. Allow us to make three brief but important stops as historical touchstones for understanding the contemporary situation. Let us look briefly at Lucretius, Galileo, and Descartes—all of whom invoked as central principles versions of the material and/or mathematical tenets so central to modern science.

Certain pre-Socratic philosophers, sometimes called "atomists," who lived between the fourth and seventh centuries B.C.E. tried to understand the natural world as a product of chance and necessity, that is, of merely naturalistic forces. Among them were Anaxagoras, Democritus, and Empedocles. Most of their texts are lost to us, and the fullest extant account of this philosophy is in Latin—Lucretius' wonderful *On the Nature of the Universe*, written in the first century B.C.E. In this work, Lucretius often gives strikingly modern-sounding accounts of atoms, infinite universes, the evolution of the cosmos, and human societies. Like many current theorists, he believed that there is no creator, no plan behind the universe, no immortal soul, no heaven or hell. For Lucretius, following his teacher Epicurus, even if the gods could be said to exist, nothing of note would follow from this because they could not affect us; we have nothing to gain nor fear from the them. As he writes in his classic poem concerning the gods:

> . . . far separate, far removed from our affairs.
> For free from every sorrow, every danger,
> Strong in their own powers, needing naught from us,
> They are not won by gifts nor touched by anger.
> Indeed the earth is now and has been always
> Devoid entirely of any kind of feeling.
> The reason why it brings forth many things
> In many ways into the light of the sun
> Is that it holds a multitude of atoms.[1]

Yet, all is not fate in Lucretius' world: individual atoms sometimes "swerve" from their original trajectories, at unpredict-

able times and places. In this way individual atoms can contact each other and combine. Certain combinations are un-
successful, but the successful ones perpetuate themselves and,
in this way, the world of our experience comes into being. As
one can see, this process is a kind of "natural selection."

Against the backdrop of this mechanistic theory
Lucretius asks the question: how should humankind act in
a universe governed only by the naturalistic forces of atoms,
chance, and necessity? Religion in the narrow sense could
not be, as we have seen, the answer for Lucretius. It is not
the answer not only because there are no gods of any consequence, but also because he feels that the religious view
of the world, with its threat of eternal torment, causes people
to fear death and to struggle for the illusory heights of
power and wealth, in an attempt to avoid the dark depths
and evils of hell. This, he sees, produces more misery than
it alleviates. Once we realize that we need not fear death,
because gods have no power over us and there is no hell, we
can adopt a larger perspective in which we see that the
countless atoms of which we are composed are involved in
an endless cycle of decomposition and recomposition. Re-
leased from our fear we can also free ourselves from endless
fruitless struggles for power, live a tranquil, philosophical
life, and perhaps more importantly we can continue to establish true friendships.

From a contemporary standpoint, we may be tempted
to say that the pre-Socratic philosophers merely made a few
observations and perhaps some lucky guesses and that their
doctrines concerning the workings of the material world
were not grounded on solid experimental, mathematical,
and theoretical foundations. This would be, however, to miss
the most salient point at issue in Lucretius' classic text and
its relation to our most pressing concerns today. In many
ways the main question raised by Lucretius is the very one
sociobiologists and other authors so often raise in our own
time: Given a universe that works only through material
causality, in which humans have no purpose other than the

reproductive dictates of their genes, how can we formulate and act according to "higher" moral principles? Not only is the question similar, but contemporary scientific theorists, after similarly arguing for a meaningless and materialistic world, also are unwilling to reside inside the worlds they have described. Somewhere in their works there is at least a nod to freedom and/or a higher moral calling.

The atomists formulated their theories and models on the basis of what we might call "idealized experience." That is, their theories were derived mainly from observations of their world together with a few first principles, and their methods were not experimental nor scientific in the modern sense. Modern science, especially physics, relies on the a priori truth of mathematics to ground its claims for knowledge. Thus, the truth of mathematical propositions is given beforehand and if we can match nature to it, then we can be confident about the soundness of our conclusions. This view has extensive roots in Western thinking, going back to Pythagoras, the sixth century B.C.E. Greek philosopher.

One of the most important heirs to and creators of this mathematizing tradition in Western thought is, of course, the Italian scientist Galileo (1564–1642), usually cited as the founding father of modern mathematical physics. A champion of the Copernican view of the solar system and discoverer of the mathematical laws that describe falling bodies, Galileo's work, like Lucretius', has contemporary resonance. Galileo used geometry to discover the laws of the physical world. He insisted in his now-famous phrase that the book of nature is written in the language of mathematics.[2] This leads him to argue that if we are to have any hope of understanding nature, it would have to be through mathematics. Furthermore, he demanded that personal observation and experiment replace the authority of the Church-sanctioned Aristotelian schoolmen on matters scientific.

While Galileo's contributions at this level would be enough to merit a place in our historical overview as a significant leap along the path pointed out by Lucretius' atomistic materialism, his work is important to another of our immediate concerns. Galileo pioneered the genre of science writing for the lay person. Writing in the Italian vernacular, rather than academic Latin, he combined masterful physical and mathematical arguments with elegant prose in his *Dialogue Concerning the Two Chief World Systems* and *Two New Sciences.* In these works, Simplicio, an Aristotelian schoolman, Sagrado, an intelligent citizen, and Salviati, the mouthpiece of Galileo, question, discuss, and explore, through observation, mathematics, and experimentation, the nature of the world.[3]

The use of mathematics as the paradigm of certainty, and the extension of this use into many branches of science and philosophy is associated especially with a contemporary of Galileo, René Descartes (1596–1650), who is widely held to be the father of modern philosophical and scientific method. His writing is subtle and complex, and it is often difficult to clearly summarize his position on many important questions. However, this much is beyond doubt, Western culture has inherited a certain version of Cartesian thought that stresses the goal of certainty and the indispensability of mathematics to guide us towards this goal. The type of certainty Descartes' philosophy seems to promise functions as an ideal in both modern science and many branches of modern philosophy. Moreover, all branches of science attempt to cash in on the certainty of mathematics by invoking its method and apparent, incontestable truth.

It is also the case that the concern for freedom and the contemporary attack on it have their precursor in Descartes. For freedom according to Descartes leads to sin and error if it does not conform to the truth, and hence is a source of danger rather than opportunity for us.[4] This is akin to the religious argument that God gave humans freedom and the greatest use of it is to conform ourselves to God's plan— to use that freedom to choose God's necessity. As we shall show in chapter three, confining freedom to an ability to recognize the given and necessary truth instead of allowing it a productive or creative power is another hallmark of modern thought.

This method has had its benefits and is a fundamental part of our Enlightenment heritage. Yet, this scientific heritage has very often suppressed the Enlightenment's commitments to the creative aspects of human freedom. We now turn to Immanuel Kant (1724–1804) to investigate these two components of the Enlightenment, for it is in the philosophy of Kant that we get a sense of what is at stake in our contemporary debates.

Kant and the Enlightenment

We have been pointing to some places in the Western tradition where the questions with which we are concerned are formulated and different positions propounded. A particularly important moment in the history of these questions is the Enlightenment. Modern science is frequently seen as a legacy of this critical period in European history when both scientific and moral questions were reconsidered in new contexts. The philosopher Kant set the stage for future philosophy with his critiques of pure reason (science), practical reason (morality), and judgment (morality, politics, and aesthetics). He believed we all needed, and indeed we are all able, to ponder three essential questions:

- What can we know?;
- What ought we to do? and;
- For what may we hope?

Kant was spurred on by the work of the Scottish philosopher David Hume (1711–1776), whose extreme skepticism about our ability to be certain about even the mere existence of the material world or natural laws led him to postulate physical and moral principles that relied only on

experience. Hume was equally suspicious of grandly conceived natural and moral systems. Humean skepticism forced
those attempting to address questions about what we can
know and what we ought to do into deeper reflection about
issues that previously seemed unproblematic. In order to
reply to the force of Hume's position, Kant reconsidered
many time-honored philosophical presuppositions.

Kant referred to himself as the "Copernicus of meta-
physics" because of the way his system of thought drastically
reorients our thinking about the world. In Kant's philosophy, instead of asking what the world is like, we are led to
inquire about the structure of our own minds and the
manner in which they orient, constrain, and determine what
we can know of the world and, consequently, what we are
prohibited from knowing.

In his first critique, *The Critique of Pure Reason*, Kant
gives us the philosophical foundations for the modern approach to the study of nature, citing the great successes of
Galileo and others in the field of physics.[5] He tells us that
we must not be passive observers of nature rather, we must
be active in determining the questions we wish nature to
answer. We must decide in advance the kind of answers we
want and then, through experimentation, cause nature to
reveal itself in ways that conform with the fixed principles
and laws that govern our understanding. Only in this way
can we gain knowledge about the physical phenomena that
constitute the proper subjects of scientific knowledge. Many
who see science as the ultimate ground of all knowledge
stop following Kant's work after these epistemological
reflections in his first *Critique*. However, Kant himself does
not stop at this point, but goes on to draw out the implications such a view of rationality has for ethical considerations.

The most provocative and important move in Kant's
philosophy is his contention that reason has limits and because of its inability rationally to transcend these limits, reason alone is incompetent to decide certain questions with
definitive empirical proof. When reason attempts to transcend

these limits, it makes claims beyond those we are able to derive from the combination of experience, experiment, and logical thinking. Kant illustrates the limits of reason with his famous antinomies, which consist of logical arguments convincingly disproving both sides of crucial scientific and moral questions. For instance, the questions of whether time is infinite or bounded and whether freedom exists or not are taken up as antinomies. Kant provides tight logical arguments denying the possibility of infinite time and others disproving the possibility of a beginning of time. He argues against the concept of freedom and, in turn, provides equally strong arguments against a totally deterministic world and hence the absence of freedom. In addressing these antinomies, he shows that possible experience coupled with logically consistent argumentation is not sufficient to decide definitively in favor of either side of these propositions.

In working out these arguments, Kant shows both sides are able to reduce the others' claims to absurdity by revealing inherent contradictions. Consequently, such claims do not, according to Kant, allow for empirical "proof" because of the contradictions one faces when following reason itself into the examination of each pair of mutually exclusive positions. While these propositions are *rationally* undecidable, it remains the case that for Kant, at least on moral issues, we nevertheless *must* decide.

Indeed, Kant claimed to have written *The Critique of Pure Reason* to set the stage for his examination of moral questions. In his first work, he argues that we can have no certain knowledge (what we called "proof" above) about God, Freedom, and Immortality, since they are outside the competence of pure reason's functions, that is, beyond empirical justification. His second major work, the *Critique of Practical Reason*, moves on to discuss the subject of ethics within this theoretical framework.[6]

If his first question "What can we know?" is answered, in part, by the response "reason has its limits," then the second question "What ought we do?" demands closer examination because the answer cannot be simply "whatever reason clearly says." Thus, in this second *Critique*, he insists that, though we lack certainty, we must still act *as if* we had the freedom to break the chain of material causality. Kant argues that if we are not free, then we are forever caught in the mechanistic world of Nature in which we can do nothing but those things dictated by natural necessity. If we cannot act but are forced only to react in ways forced upon us by necessity, then the very concepts of ethics and responsibility make no rational sense. Consequently, for Kant, ethics only exist when we believe we could have acted otherwise. For Kant, ethical concerns arise when we take responsibility for our actions and believe that we can act for the dignity of our souls, according to a sense of duty to ourselves and our fellow human beings. Unlike most of his predecessors, Kant believed it was vital that all humans could and ought to learn philosophy and that they learn to think and judge for themselves. Only then, he thinks, could the human race as a whole progress and universal values be established. Thus, Kant shows that we decide in favor of freedom, not because of empirical verifiability, but rather, for ethical reasons.

As we saw earlier and will see again, those promoting a scientific world view today often see themselves as the only legitimate heirs of the Enlightenment, fighting against the evils of irrationality, be it in the form of a belief in free will, religion, or astrology. However, in his essay "What is Enlightenment?" Kant argues that the essence of Enlightenment is when each person thinks and speaks publicly for himself and does not submit to authority in matters of belief, whether secular, scientific, or religious. Today, whether or not it is stated explicitly, science and its methods are presented as the only rational alternative available to humankind to overcome the evils of religion, ideology, and superstition. From this viewpoint, anything not scientific is irrational at worst and a-rational at best.

What makes Kant so important to our thinking is that, great admirer of science though he was, he was also able to

see the world *otherwise* than through the powerful but limited
view of scientific rationality. He argued strongly for the place
of science within rationality, but he also argued that it was
equally rational to keep science and its methods from overtak-
ing every domain of thinking. Kant was convinced that science should respect the limits within which it has its
effectiveness. Beyond these limits of scientific thinking lie the
realms of faith and freedom. This nonscientific side of Kant's
thinking is an important legacy of the Enlightenment in its
own right, even though (and we would say more so because)
it is all too often pushed into the shadows by scientism.

Both the scientific and ethical aspects of Kant's philosophy recognize the importance of progress—scientific
progress in understanding phenomena of the material world
and moral and political progress that would lead us toward
an era when humans would follow moral imperatives toward
what he called a "free and cosmopolitan organization of society." Rationality is not exiled from the realm of morality,
but can only function for the betterment of humanity after
we accept the rationally unprovable position that we are free
and have moral obligations. Then, and for Kant only then,
can we use our reason to act in accordance with this belief.

The Emergence of Evolutionary Theory

Through this admittedly brief and selective history—from
the dawn of Western thinking to the Enlightenment—we
can see that questions concerning knowledge of the material world and its connection with human nature and the
relation of freedom to moral conduct have been the subject
of much tension and debate. We feel this history is important (and other stops along the way could tell a similar
story) because it informs these same debates in the various
guises in which they continue in our own time. We have
focused on two major themes: we saw that the quest for
certainty has a relation to questions in both the material
and moral realms and, at some level, all of these questions

are related to the necessity of deciding whether freedom exists and if it does, the extent of its force.

These issues take on the forms which are most readily recognizable in contemporary debates with the publication of Charles Darwin's *Origin of Species* in 1859 and *Descent of Man* following in 1871. These books gave evolutionary biology its distinctively modern form and revolutionized the study of all biology by giving it a general framework in which everything seemed to fit. Since Darwin's time, it is undeniable that this theoretical framework enormously and fruitfully guides biological research. These classic works also begin the modern debate on the role of natural selection in determining the behavior and emotions, as well as the intellectual and spiritual life of humankind.

As one would expect and as historians and rhetoricians have documented, Darwin's books immediately led to protests by religious communities. They were not alone, as objections were also raised by the learned community: philosophers, men of letters, as well as many scientists of the day. In fact, the objections raised in the years following publication of these books are essentially the same as the objections we hear today, despite the fact that our century has seen unparalleled advances in our understanding of the dynamics of populations and the mechanisms of inheritance. We might even say that despite what the scientific community might reasonably expect or wish, namely, that the success of the Darwinian explanation would quell such debate, it seems to have had an inverse consequence. Indeed, the more compelling scientists judge the case to be, the more voices there are wishing to join the dissent. Though objections to Darwinian theory are often couched in terms of the details and adequacy of the theory vis-à-vis objective data, the persistence of the same basic objections—on shifting 'scientific' grounds—should make us pause and ask if something more is really at stake than the facts of evolution.

The phenomenon of the resistance to modern science from such a great variety of sources religious, intellectual, and lay should lead us to ask some questions as to why an endeavor, which on the surface promises so much by way of knowledge of the world, utility to humankind, and a way out of superstition and the evils of religion, can still be so widely suspected and rejected. As must be clear at this point we suspect the reasons are not merely ignorance, superstition, nor evil.

Despite the advances of science we still find critiques continuing to come from religious thinkers. Yet, we must note not all religious critiques are the same. For example, Christian Fundamentalism, in the guise of creation science, has been particularly vocal in criticizing mainstream evolutionary biology—and it frequently appears as the enemy to be defeated in books and articles about evolution or modern science. These critiques are often rejected simply because they are voiced by religiously motivated individuals. Indeed, these creation science positions are suspect because they are clearly advanced with the agenda of discrediting conventional science so that the current scientific account of evolution can be replaced with the biblical account of the Creation. The majority of these arguments imply that the discrediting of current scientific views necessarily leaves only the Christian God of the New Testament left as the explanatory cause and meaning of life.

Although we do not support this agenda—indeed we find it frightening for a number of reasons that can be documented by reviewing their literature—we do feel that the questions raised by religious thinkers, if based on questions of scientific evidence for scientific claims or the illegitimate extension of these scientific claims to religious questions, should be given all the attention they merit.[7] Because so many today believe that the debate is solely between creationism and evolutionary theory, an example of a thoughtful critique generated from a religious perspective is worth considering. Too often we find works from these perspectives dismissed because their religious motivations are used as sufficient grounds to ignore the relevant arguments they may make.

A fine example of a religiously motivated, though not fundamentalist, critique is Phillip Johnson's book *Darwin on Trial*.[8] While we do not agree with all of Johnson's positions, he ably summarizes the evidence (or lack thereof) for current theories about the origins of life and macroevolutionary change (changes which seem to have occurred in large, discrete steps, rather than gradually). He brings a lawyer's acuity to the rhetoric used by evolutionary biologists to defend their beliefs, as well as assesses what he sees as their hidden, antireligious agenda.

Johnson, as well as others, stresses that the statements of many present-day evolutionary biologists, while plausible, reach far beyond the scientific data. Scientists making such claims and advancing these positions offer us a powerful world view in which science is able to take on all the functions previously served by religion. Here, however, we notice a striking similarity between scientists' accounts of life, in the broadest sense, and the religious accounts they so seek to discredit. Like the Bible, the writings of these scientists offer us a world view that includes a great creation myth, be it primeval soup, life from space, or aggregates of molecules on clay, which explains the origins of living from nonliving matter. It also accounts for the "evil" nature of humankind through the necessary selfish imperatives of our genes. Scientists frequently point to the promise of eventual salvation through scientific understanding and technology.[9] From this perspective, as Nietzsche in the next chapter will help us understand, evolutionary biologists and other scientists appear in the rather surprising theoretical company of those they have traditionally reviled—speaking oracularly to us of origins, good and evil, and salvation.

The Promotion of Human Sociobiology

The wish to understand the essence of human nature, both intellectually and as a guide to action, seems nearly ubiquitous and occupies our thoughts at least at some point in our lives. There is a substantial public appetite for popular books and articles on the subject, including many that look at the question from a scientific point of view. Most, if not all, of us are familiar with such books. We have read many of them and they frequently receive high praise in the popular press and are the subject of articles in general interest and news magazines as well as academic journals.

At the time of this writing, as there would have been at any other time in the last 60–70 years, there are a wide variety of examples from which to choose. To set the stage for our discussion, we begin with a few telling examples chosen for their ability to demonstrate not only current popular scientific journalism addressing these issues, but also for their ability to demonstrate the variety of forms and forums in which such ideas are presented.

In a striking example of this interest in scientific accounts of human nature, *Time* magazine ran a cover which proclaimed in large letters "Infidelity: It May Be In Our Genes." The article by the journalist Robert Wright is entitled "Our Cheating Hearts."[10] This article champions the cause of the science of evolutionary psychology, which claims that Darwinian theory finally gives us a well-grounded scientific account of human nature. The same author has found other venues, publishing numerous articles in *The New Republic* and *The New Yorker*.[11] However, as we have already noted, these pieces do not stand alone, nor are they particularly unique in their thesis. The scientific understanding of human nature that these articles advocate is typical of many books written on the subject by both scientists and journalists.[12]

As with any set of ideas there are surely individual differences among these thinkers, though the premise underlying all their analyses is rather simple and straightforward: The social, emotional, and intellectual life of animals—of which the human species is but one example—are to be understood just like all other corporeal and behavioral characteristics and are best explained, in the end, through the principle of natural selection. If we begin by recognizing that many more offspring are produced than can survive given limitations in resources such as food and nesting sites, then any organisms whose chance genetic mutations give them even the slightest advantage over others, and which are thereby enabled to produce more offspring that survive to reproductive age, will prevail. In a distinctive modern form, this purely mechanistic model of evolution stresses the survival, not of individual organisms or species, but of the genes which they carry. For instance, Wright, an advocate for this view, writes: "The human mind, was designed, like any other organ, for the purpose of transmitting genes to the next generation; the feelings and thoughts it produces are best understood in these terms."[13] According to this sociobiological theory, human beings, under the influence of genes that evolved in prehistory, are prone to act in such a way that their number of offspring living to reproductive age is maximized.

Given these premises, we should expect the psychology and behavior of men and women to be very different and thus we have a "true" explanation of the oft-noted differences between men and women. A popular version of the argument goes like this: since men can theoretically produce many more offspring in a lifetime than can women, the optimal strategies for assuring maximal representation of genes in the next generation should differ for the two sexes. Such a premise leads to claims that on the one hand men should be more promiscuous, seeking out and accepting many sex partners and on the other hand women, who must bear the burdens and dangers of pregnancy, childbirth, and nursing, should seek men who will stay around and help them raise more offspring by contributing financial and other forms of support. Man should be designed, above all, to guard against cuckoldry, against raising other men's offspring, and his emotions and actions should reflect these "innate" motivations. Similarly, woman should be most concerned with maintaining the support of a man. Of course, it may be advantageous to her to trick a man with "worse genes" but more money into thinking he is the father of her child. She may also cheat to make other men think they have, or may have, fathered her offspring with the result that they may give her gifts or merely refrain from hurting or killing her children.

We are told by proponents of this position: "Men and women have always, in a sense, been designed to make each other miserable, but these days they are especially good at it."[14] The various emotions and actions that lead to the breakup of marriages and our society's high divorce rate are simply our genes' way of making more copies of themselves. We are told: "Beneath the thoughts and feelings and temperamental differences marriage counselors spend their time sensitively assessing are the stratagems of the genes—cold, hard equations composed of simple variables: social status, age of spouse, number of children, their ages, outside romantic opportunities and so on."[15] Though it is difficult to evaluate a claim like this, we might wonder just how "simple" concepts such as social status are, and what it means to reduce them to a number in an equation.

Many of these writers accept uncritically that evolutionary theory, including its specifically sociobiological aspects, is sound. The purpose of these books and articles is to show how evolutionary psychology leads to a better understanding of human behavior and at last sets the study of human psychology on "something firm: an understanding of the process that designed the human brain."[16]

The relevance of such a theoretical commitment to sociobiology becomes clear to us if we look at the implications that arise when these theories are used to interpret social behavior. We can start to draw out some of these implications with a look at the interpretation given to monogamy. The interpretation begins by stating that monogamy is "unnatural" for both sexes according to sociobiological theory, but it is then described as being good for the welfare of children according to sociological sur-

veys. If we assume that one of a society's goals is to promote the welfare of children, a solution offered from the
viewpoint of evolutionary psychology is to invoke what
must presumably be "unnatural" political mechanisms such
as, in their words, "soaking the rich" so as to distribute
income more evenly and thus prevent men from garner-
ing an unfair number of reproductive women. These not-
so-wealthy men would now presumably (though we note
these motivations are never explained) stay home faith-
fully with their wives and help to support their children
monetarily and otherwise. Even if we accept the claims of
evolutionary theory, we may be dubious that such complex sociological problems are best analyzed in terms of
a simple biological model.

As these examples suggest, and again there exist countless others, all aspects of our lives must be scrutinized through
this sociobiological lens. For instance, in *The Moral Animal*,
Wright notes that "The Darwinian line on spiritual discourse
is much like the Darwinian line on moral discourse. People
tend to say and believe things that are in their evolutionarily
ingrained interests. This doesn't mean that harboring these
ideas always gets their genes spread."[17] In this context, Wright
mentions celibacy as a religious doctrine that clearly fails to
spread genes, but does not attempt to give an account of
this empirical fact, a fact moreover at odds with the broad
theoretical commitments that are used to generate evolutionary descriptions of social practices. As we shall see when
we look at Darwin's work, Darwin himself was troubled by
this same phenomenon. The choice of a celibate life certainly appears to run contrary to the expectations of evolutionary theory. Darwin, contrary to many who write in his
name today, engaged the issue with no small amount of
intellectual care and honesty.

The handling of this issue may well be enough to raise
our intellectual suspicion, yet it is not only cases of social
behavior that become the objects of evolutionary explanation; long-standing philosophical beliefs come under attack

from the camps of sociobiology as well. A particularly difficult problem for this group of theorists is the widely held belief in freedom and free will. Free will (broadly considered as the ability to choose freely between alternative courses of action), in the sociobiological view, is a delusion. Staying with Wright as our example: in a rather convoluted explanation, Wright contends that this delusion, brought about by evolution, is an adaptation to hide our real motives for action not only from others but even from ourselves. These motives are hidden from us not incidentally, but by design so that we can credibly act as if they are not what they "in fact" are. Wright admits that the delusion of free will may have been useful in the past. But now, with the knowledge gained through evolutionary science, "free will is beginning to outlive its usefulness. After another decade or two of biological research, it may be more trouble than it's worth."[18]

We are troubled by the vocabulary to which such a position leads. Such theorists argue that we are "robots" determined by a combination of genes and environment, but insist as well that it is possible to make robots responsible for their malfunctions. They contend that we have the ability to resist the "natural" impulses of our genes, and, by pondering Darwinism, choose to live by seemingly non-Darwinian values such as universal compassion for fellow humans. We wonder whence these others values and what would motivate us to choose and embrace them? Does not the idea of choosing them suggest a sense of deliberation by free beings that might do otherwise? Indeed, how could we come to think of them, consider them, or choose to live by them if we are but reacting machines in determined material circumstances?

As we have noted, such thinking comes from many sources. Richard Dawkins and Daniel Dennet are two prominent representatives of this brand of "ultra-Darwinism." They believe that all biological and cultural phenomena are explained by "Darwinian" principles. The possibility of combating the selfish dictates of our genes is addressed by Dawkins and by Dennett through their notion of "memes" which are to be taken as the cultural equivalent of biological genes. Memes are attractive, "addictive" patterns of thought or culture whose interest is to perpetuate by spreading themselves in a human group (they include such things as advertising jingles, which we may dislike but which stay with us anyway). They are "selfish," that is, they need not be biologically adaptive for us, their carriers. For Daniel Dennett and Susan Blackmore, two proponents of this theory, memes take on a relative autonomy from the dictates of genes though they function in a purely mechanistic way. Blackmore goes beyond even Dennet when she claims in her chapter on the illusion of free will that ". . . free will is just a story that forms a part of a vast memeplex, and a false story at that."[19]

The moral conclusions of these writers, though they rely on arguments from our inherent selfishness, are not especially heinous, nor do they advocate a selfish social Darwinism in which whoever wins is right. Others using the same sort of argumentation have reached much more pessimistic or morally repugnant conclusions.[20] However, when we see that from essentially identical theoretical positions vastly different moral views are promulgated, we should be cautious about resting any moral conclusions on the findings of evolutionary biology.

What is really at stake here? Can we determine what is worth living for and live according to these ideals if they run contrary to the predictions of evolutionary theory—either in a material/reproductive sense or in the broader cultural sense of memes? Any attempt to answer these questions must take seriously the issues that have surrounded the age-old dilemma of determinism and free will. We will focus many of our philosophical concerns around this issue and, to take this old debate into the arena of our contemporary world, we now suggest some questions that are relevant to such a discussion. Are the arguments advanced by proponents of evolutionary theory against the existence of free will adequate? Further, and perhaps more important, if they could be shown to be decisive, would this be grounds for banishing such a seemingly life-enhancing concept from our culture? Surely our attitudes toward persons and situations are derived, in part, from what we believe concerning questions of determinism and free will.

The passion of these defenses of the scientific view of the world suggests to us that there is more at stake than questions of scientific evidence. People throughout the ages have observed and theorized about the natural world and have held a range of theoretical positions that bolstered and guided their inquiries. Further, as we have shown, these basic philosophical positions were elaborated millennia before any of the modern scientific evidence that is seen as giving some positions legitimacy and discrediting others was available. As we saw in Lucretius, atomists believed that chance and necessity guided the universe; Platonists espoused a theory of real forms or mathematical concepts that gave the universe its truth; deists gave credit to a supernatural intelligence; Aristotelians saw a vital power at work in animating life; there have been no shortage of attempts to develop explanatory theories for matter, life, and their relation. Those who come to the study of nature with prior convictions, about the existence or nonexistence of God or some vital force that operates on matter, can find evidence and construct arguments that support their views and weaken or refute the views of others.

We may need to ask ourselves what our ultimate aims are. If we want to believe in and work toward moral and social progress, a purely mechanistic view of nature seems opposed to these aims, and we must look for evidence of freedom or another ultimate purpose toward which we may put our efforts. A grossly reductionistic account misses much of what we see around us as well as the "truth" of our own inner experience. Given this diversity, and, if we follow Kant, the impossibility of certainty on these matters, why should it matter which views we hold? Why should the scientific world view dictate what may be thought of as the truth about nature? It is apparent from reading many of the books addressed in the last few pages that it mattered greatly to the people who poured their passion, energy, and intellect into their work and writing. It mattered to them, and they wanted others to be convinced of the truth of their findings and of the implications these findings suggested to them.

It is odd that even with the astonishing scientific advantages of our century and the increasing sophistication of our theories these lasting questions about the essence of nature and the existence of God or free will still remain, suggesting that the same "wooden-headed" objections scientists have thought they had put to rest once and for all are inconceivably back again. We say odd because the persistence of these questions might speak not of the stubbornness, superstition, and stupidity of those who raise them, but, to the contrary, to their perennial importance. Indeed ancient mechanistic explanations of the world were not disputed by Platonism or Christianity simply because we did not have enough facts yet or had not discovered the laws of physics but because they fail, for many people, to give a satisfactory answer to many central questions in life—and even to prohibit questions that seem vital.

Many scientists will admit there are questions science cannot answer, at least not yet. However, and this goes to the heart of our concerns, many scientific writings make this admission, only to fall quickly back into the use of the "facts of objective science" to answer these very questions which "for the record" they have claimed to be currently unanswerable. Richard Dawkins is a master player of this double game. He cites G. G. Simpson on page one of *The Selfish Gene,* and whom we cite at length:

> The question "What is man?" is probably the most profound that can be asked by man. It has always been central to any system of philosophy or of theology. We know that it was asked by the most learned humans 2000 years ago, and it is just

possible that it was being asked by the most brilliant australopithecines 2 million years ago. The
point I want to make now is that all attempts to
answer that question before 1859 are worthless
and that we will be better off if we ignore them
completely.[21]

In his endnotes to the book Dawkins says the following about
the above quotation:

Some people, even non-religious people, have
taken offence at the quotation from Simpson. I
agree that, when you first read it, it sounds terri-
bly philistine and gauche and intolerant, a bit like
Henry Ford's "History is more or less bunk." But,
religious answers apart (I am familiar with them,
save your stamp), when you are actually challenged
to think of pre-Darwinian answers to the question
"What is man?" "Is there a meaning to life?" "What
are we for?", can you, as a matter of fact, think of
any that are not now worthless except for their
(considerable) historic interest? There is such a
thing as being just plain wrong, and that is what,
before 1859, all answers to those questions were.[22]

As if this were not enough, let us look to the following from
Dawkin's popular *The Selfish Gene*:

Be warned that if you wish, as I do, to build a
society in which individuals cooperate generously
and unselfishly towards a common good, you can
expect little help from biological nature. Let us try
to teach generosity and altruism, because we are
born selfish. Let us understand what our own selfish
genes are up to, because we may then at least have
the chance to upset their designs, something that
no other species has ever aspired to do.[23]

This book by Dawkins and his most recent collection of essays go on for hundreds of pages, teaching us that we are mere vehicles for DNA with no other purpose than to allow our DNA to replicate itself. Then a few caveats like this are thrown in, as mere assertions, which in fact contradict the argument of the hundreds of other pages. If there is only one principle to man, and that is evolution by natural selection, then no amount of teaching could challenge this ultimate principle. If teaching can make us cooperative and able to work toward the common good, then humans must be more than just vehicles for replicating selfish genes and we can happily bring back under consideration the many thousands of writers, both religious and secular, who have thought and acted on some other principle toward these cooperative ends. If evolutionary thinking turns out to have a useful part in these considerations, then we can read those books too.

It is from this perspective and against the background sketched in this chapter that we pursue our inquiry of moral evolution. Nietzsche's *Genealogy of Morals* was published sixteen years after Darwin's *Descent of Man*. Starting from many of the same assumptions, these authors give provocative alternative accounts of the evolution of moral values. In order to trace the lineage of the contemporary debate, we begin not with the Darwin of Darwinism nor the Nietzsche commonly associated with fascism and a bleak and nihilistic world view, but with each of their arguments in their own terms.

Chapter Three

What Darwin and Nietzsche Saw

The Vitalist philosophers made no such mistakes.
Nietzsche, for example, thinking out the great central truth of the Will to Power instead of cutting off
mouse-tails, had no difficulty in concluding that the
final objective of this Will to Power was power over
self, and that the seekers after power over others
and material possessions were on a false scent.

—George Bernard Shaw,
preface to *Back to Methuselah*

Some Background on Darwin and Nietzsche

Charles Darwin's enduring contributions to science and
natural history are astonishing in their scope and detail. He
is best known as an evolutionist whose book *The Origin of
Species* revolutionized the way biologists interpreted the natural world.[1] He is less well known for his extensive and
magnificent monographs on such subjects as earthworms,
orchids, barnacles, and carnivorous plants. As readers of
these works will find, his love and delight in the natural
world are infectious. His powers of observation, wealth of

45

detail collected from other observers, and careful reporting
and analysis of problems set a standard which very few biologists can hope to equal. Darwin was capable of both
grand theories like that in *Origin of Species* and the minute
attention to details and recognition of the small differences
among individual organisms that we find in his monographs.

The theory advanced in *Origin of Species* has been the
subject of continual controversy since its publication attract-
ing, as we have seen, equally passionate detractors and defenders. The position Darwin propounded in this book is
distinctly pluralistic. He certainly stresses natural selection
as the material cause by which species are modified and
eventually changed into new species. This aspect of his work
is the one most often cited and exploited in contemporary
appeals to his thought. However, and this is often obscured
in contemporary discussions, he also invokes many other
causes alongside natural selection. These include habit, use
and disuse, judgment, and various unknown laws of growth
and variation. Even "Lamarckian" factors, by which learned
habits can become fixed through inheritance, factors many
believe Darwin rejected, are invoked, though they are not
given center stage.

Today's biologists are apt to claim that our current
theories and knowledge of genetics allow us to do away with
extraneous causes, that these nonmaterial factors invoked
by Darwin himself have been subsumed under natural selection or other chance mechanisms. Contemporary authors
committed to the evolutionary stance, such as those discussed in the first chapter, seem quite sure that Darwin
would have been pleased, for example, with the theory of
kin selection that attributes all our seemingly altruistic traits
to an underlying selfishness. Though recognizing the possibility that all could be due to chance and selfishness, Darwin is not inclined to attribute our highest faculties to these
causes and takes pains in his work to argue against this
position. He offers us instead a more complicated vision in
which selfishness, reason, social instincts, and religion all

play a role. Thus, says Darwin "the reproach is removed of laying the foundation of the noblest part of our nature in the base principle of selfishness."[2] Perhaps the Darwin of *Origin of Species* and *Descent of Man* is not the Darwin so often defended and attacked in the name of Darwinism.[3]

Darwin came to the problem of the moral and intellectual faculties of humankind from his evolutionary perspective. The book with which we are chiefly concerned here, *The Descent of Man,* is his attempt to analyze these difficult problems from the standpoint of natural history. Darwin believed that for evolutionary theory, the high standard of our intellectual powers and moral disposition is the greatest difficulty that presents itself. In this Darwin was not alone, for from a genealogical perspective, Nietzsche understood this to be the greatest difficulty as well.

While they shared this concern for the evolution of our moral and intellectual powers, Nietzsche's training and intellectual life was very different from Darwin's. He studied intensely and then briefly taught philology, and had a special love for the great Greek tragedies. His first book, *The Birth of Tragedy,* revolutionized our understanding of what tragedy meant for the Greeks and introduced one of his lifelong obsessions—the tension between the rational and the irrational in our lives. What Nietzsche may well have learned from these tragedies is that the irrational is suppressed at the peril of losing *everything* we value. Tragedies such as Euripides' *Bacchae, Medea,* and *Hippolytus,* and Sophocles' *Antigone* all deal with this issue and show the destruction of individuals, families, and kingdoms that results from an overvaluation of our rational minds and a failure to appreciate and give the irrational its due. Nietzsche suggests in this book that reason's foremost irrationality is its attempt to eradicate once and for all the whole of irrationality and desire.

Ill health forced Nietzsche to spend much of his life in seclusion where he wrote the books that would radically challenge, if not completely change, philosophy. Unlike

Darwin whose work sent shock waves across the intellectual landscape of the nineteenth century, Nietzsche did not receive recognition for his achievements during his lifetime. The lack of acceptance or even knowledge of his ideas did not deter Nietzsche, for he devoted himself to trying to understand what motivated individuals, peoples, and whole nations. He re-evaluated in shocking ways the values we hold most dear and many of the truths that seemed, and may still seem, the most evident. The book with which we will concern ourselves here, *The Genealogy of Morals*, provides an entirely new assessment of the development of Western civilization—concerning itself particularly with the descent of civilization from the Ancient Greeks through Christianity, up to the modern Western scientific view of the world.

Nietzsche is a notoriously difficult philosopher to read well. He usually wrote in an aphoristic form that is cryptic and dense. As he himself warns, to read such works well one needs much time for rumination, reaction, and digestion. The *Genealogy of Morals* is probably his most accessible book in that the arguments are his most sustained and take the form with which we are most familiar. Despite this denseness, Nietzsche's work in all its articulations has a sense of coherence. To illustrate, Nietzsche wrote the *Genealogy* to help explain an earlier book, *Beyond Good and Evil*, which was itself written to elucidate a yet earlier work, *Thus Spoke Zarathustra*.[4] In contrast to Darwin's rather staid prose, Nietzsche writes to shock us into actively considering areas of our lives that we are inclined to take for granted, those areas, as Nietzsche points out, we may rather wish to leave unexamined. As comfortable as it might have been to leave these areas unexplored, Nietzsche thought that to do so would be a great danger. He foresaw a serious crisis ahead for European civilization, including the attempted destruction of European Jewry, and felt that attention had to be called to our most cherished presuppositions and truths. He was committed to examining these questions from every possible intellectual angle, despite the emotional consequences.

Nietzsche knew he would offend and be misunderstood. And this has indeed occurred, partly because of the unusual quality of his writings—not dry, logical arguments but rather poetic, humorous, and charged with meaning—and partly because most people do not put in the necessary time to read him closely and patiently. Like all great thinkers, Darwin of course included, Nietzsche's work ought not be reduced to simple slogans.

Darwin's and Nietzsche's examinations of the origins of our moral and intellectual qualities challenged previous thinking. Many people found, and still find, their accounts disturbing. However, the work of these authors is all too often familiar only through hearsay and their positions are all too frequently misrepresented—certainly by their detractors and, ironically, often by their defenders as well.

The two works we will be discussing here are especially deserving of our attention because, like modern sociobiology, they attempt to give *naturalistic* accounts of the development of humankind's intellectual and moral faculties. Thus, neither author invokes an omnipotent, Creator God, though both do try to analyze the idea of God and the utility of this idea for humankind. Both use facts as well as an historical perspective to try to understand how we developed our moral values and what the future might hold for these values and the people who practice them. They also share a surprising number of presuppositions and intellectual tactics. Despite these similarities, however, their accounts and conclusions diverge widely and it is a version of this difference that we have inherited today.

The key to understanding their divergent assessments of human moral evolution is the problem of our "animal" instincts. While, as we have said, their conclusions are quite different, both start with the assumption that modern Europeans are like domesticated animals who must keep their animal instincts under control in order to enjoy the benefits of civilization. For example, Darwin speaks of "civilized men, who in one sense are highly domesticated. . . ." and elsewhere speaks of the domestication of civilized races as being incomplete because they are "not subjected to methodical enough selection" and mentions that civilized races are like domestic animals in their ability to resist changes of condition better than savages.[5] For his part, Nietzsche remarks that the "true" meaning of all culture "is the reduction of the beast of prey 'man' to a tame and civilized animal, a *domestic* animal."[6] Both also agree that we are profoundly miserable when we are prevented from expressing an instinct. Darwin writes of the feeling of dissatisfaction and misery that is the "invariable result from any unsatisfied instinct"[7] and Nietzsche refers often, in the *Genealogy* and elsewhere, to the profound physiological misery we suffer when prevented from expressing our animal instincts. In the following pages, we examine the ways these two thinkers play out the details of these rather provocative claims.

Darwin's *Descent of Man*: The Evolution of Morality

As mentioned earlier, Darwin explicitly avoids the kind of kin selection model favored by sociobiologists. He feels it a reproach to "lay the foundation of the noblest part of our nature in the base principle of selfishness" and instead lays the foundation in sympathy.[8] He defines sympathy as a social instinct that causes us to regard the disapprobation of the social group. When we are denied the approval of our social group, Darwin argues, we become miserable and seek to behave differently in order to regain it.

According to Darwin's theory two opposing tendencies have evolved. The selfish instincts would motivate a man to try to maximize his number of offspring by pursuing, let us say, adulterous relationships, while the social instincts would make him feel the greatest sense of dissatisfaction with himself at having transgressed the norms of his social group. The sexual urge is quickly satisfied, but the knowledge of his group's disapprobation can stay with him for decades afterwards. What is most important here is that these feelings of guilt and dissatisfaction are present even in the absence of knowledge of the transgression by others. For Darwin, conscience is the resolve to act differently in the future after feeling the sense of dissatisfaction that follows from acting on the immediate impulse of, for example, sexual satisfaction. These feelings arise when we do not choose to rely on the more enduring social instincts that tend toward maintaining the cohesiveness and stability of the group.

We need not, according to Darwin, be caught in the act of violating social norms because in our internalization of social norms, to use contemporary language, we catch ourselves in the act. Every recollection of such an event produces painful feelings and acts to dissuade us from trying such adventures in the future. Notice that it is not necessarily the direct physical or social consequences that the group can inflict when a transgression is discovered that motivates actions here. According to Darwin, when our evolved social instincts are thwarted, we feel a natural misery and dissatisfaction with ourselves that alone may be sufficient to urge us towards virtuous conduct. In Darwin's account, these social instincts have evolved through natural selection because those who had them were able to live more harmoniously as a group than those who lacked such instincts, thus enjoying group benefits such as mutual protection and food acquisition.

Thus, Darwin's analysis pits selfish "animal" instincts such as lust, hunger, and avarice against the more "enduring" social instincts such as fidelity, sympathy, and temperance. He believes that these social instincts have evolved for the good of the group and that groups in which they were established by natural selection and strengthened by habit and custom survived better than those in which *only* selfish instincts prevailed. In his analysis, the evolution of reason aids us in understanding the superiority of these social instincts for the good of the group, and thus aids their establishment in our communities.

Darwin saw many processes at work in the evolution of our moral sense. An account of these looks something like this: First was the establishment of the social instincts (e.g., compassion) through natural selection; second was the establishment of the virtues derived from these instincts as norms of social behavior in communities. Thus, compassion can aid the development of self command; we may refrain from stealing food because it would deprive another of sustenance. A third possibility Darwin sees is the establishment of virtuous traits through a kind of Lamarckian inheritance. Thus, virtues such as temperance and pity may first be established through custom, but later be inherited by the children of those who possess them. Finally, Darwin sees that when inheritance and reason fail, society enforces its norms through a fear of punishment and knowledge of the negative consequences sure to follow any discovery of transgressions of social norms. Even this, according to Darwin, would not be enough to lead us to the point where we could "love our enemies." In order to reach this higher order of morality, Darwin points to the need to extend our social instincts with the aid of reason, instruction, and the love and/or fear of God. Here, according to this account, is religion's value for us: The idea of God is needed if we are to reach the highest moral stage. In Darwin's telling of this story, this stage especially characterizes civilized man. Indeed, Darwin quotes Kant in reference to this highest moral stage: "I will not in my own person violate the dignity of humanity" and remarks that only cultivated man, not savages, could think this.[9]

Near the beginning of the *Descent of Man*, Darwin gives us the anatomical evidence for man's relationship to apes. In one sense this evidence is decisive, though for Darwin anatomical evidence alone lacked the rhetorical appeal necessary to convince us that our intellectual and moral qualities must also have evolved. Darwin had to go further and show how our intellectual and moral qualities could have evolved from those of animals and give us a way to feel proud of, rather than repulsed by, our origins.

To this end, the distinction "tame" versus "wild" becomes important for his argument. Darwin gives examples of the superiority of tame animals over their wild counterparts citing improvements in affection, trustworthiness, temper, and general intelligence. He gives examples of powers of abstraction and self-consciousness in other mammals and even claims that dogs regard their masters almost as men regard God. Darwin feels compelled to show that, by way of evolution, social instincts have evolved that give us the possibility of becoming tamed. Thus, he dwells upon what he sees as the deficiencies in these powers among the "wild" and "savage" nations. From this it follows for him that "tamed" animals can be morally compared with and perhaps even be considered superior in some respects to "wild" men.

For instance, Darwin dwells on the deficiency of Christian virtues such as chastity, humility, and temperance amongst "savage" nations. Darwin notes that for savages it is a reproach *not* to avenge oneself. Ever the Protestant English gentleman, Darwin sees savages as lacking the blessings of a developed reasoning power which could aid in discerning their true interests and in ridding themselves of foolish and dangerous superstitions. According to Darwin, they cannot conceive a higher morality that appeals to higher moral goods than personal and group selfishness.

If we follow Darwin's account, we see that many factors have contributed to the development of human morality. Here is Darwin's final assessment of the factors responsible for the development of humankind's higher moral nature:

> Important as the struggle for existence has been and even still is—yet as far as the highest part of man's nature is concerned there are other agencies more important. For the moral qualities are advanced, either directly or indirectly, much more through the effects of habit, the reasoning powers, instruction, religion, etc., than through natural selection.[10]

Always conscientious, Darwin did not shrink from noting some disturbing consequences and inconsistencies in
his position. He wonders, for example, if these civilizing
instincts will prevent the strongest from producing the most
offspring and whether our pity for the unfortunate will
cause our species to weaken. Darwin admits the inherent
conflict between virtues such as prudence and sympathy
for all fellow persons and the necessity of producing the
greatest number of offspring. This is a conflict, of course,
because of Darwin's belief that intense competition to
produce the most offspring causes individuals to become
ever more fit. The social instincts are caught in a tension
between what is good for the group and what is good for
the individual.

Showing an admirable intellectual honesty and curiosity, he wonders further why the most intelligent and vigor-
ous men often choose to be celibate clergy who, through
institutional norms, prohibit themselves from reproducing.
Would it not be these very men, Darwin asks, who would
contribute most to the new generation by passing on their
natural endowments? He notes that the Church itself historically executed thousands of the finest men—and yet,
paradoxically, Darwin notes, Europe had progressed at an
unparalleled rate. How are we to understand these seeming
contradictions?

When Darwin looks towards the future, he is, nevertheless, optimistic: ". . . virtuous habits will grow stronger, becoming perhaps fixed by inheritance. In this case the struggle
between our higher and lower impulses will be less severe,
and virtue will be triumphant."[11] Darwin tells us that humankind should "feel proud of his origins, and hopeful" for
having risen, though not through his own exertions, "to the
very summit of the organic scale." This rise from lowly origins gives us hope for "a still higher destiny in the distant
future." This tension, if not outright contradiction, between
passivity ("not through his own exertions") and activity ("reason, habit, instruction, and religion" as more important to

the outcome of moral development) is a curious and notable component of Darwin's theory.

On this reading then, we see that Darwin might not have been eager to accept the current sociobiological understanding of human morality. Notice also that he cites both moral and evidential reasons for rejecting such a model. He asserts that selfishness is a "base" principle and could not underlie the height of human moral attainment, and he cites the evidence of our consciences to support his belief. Modern scientists often claim that science advanced precisely when we gave up endowing nature with an active or moral dimension and stopped believing that morality and objective knowledge must somehow be compatible. Indeed, the history of science offers us numerous examples in which important advances were only made when we relinquished our time-honored notions of morality or common sense and followed a theoretical path that was not immediately suggested by experience or reason. Think, for example, of the change from an Earth-centered to a sun-centered view of the solar system. The Copernican view succeeded only when we gave up our reliance on experience (we see the sun "rise" and we do not feel the Earth turning) to guide our theories and were able to let go of the necessity for complete compatibility between our mechanistic accounts of the heavens and an official interpretation of the Holy Scripture.

Given such scientific precedents as are exemplified by Darwin's brilliant work, it may indeed seem difficult to cling to our conventional understanding of morality instead of following the road that the sociobiologists have, in Darwin's name, pointed out to us. It is possible that the critiques of current sociobiological theory will turn out to be fundamentally misguided and that the social, intellectual, and moral qualities of our species will be ever more convincingly illuminated as we follow this path. The sociobiological account does seem to give satisfying explanations to certain widespread, if not ubiquitous, phenomena in human and animal communities. For instance, nepotism is a well-attested human trait, as are widespread double standards of male and female sexual behaviors. Evidence from other animals is cited when such studies seem to support the sociobiological analysis of these behaviors in humans. If it were not for such evidence and the power of its prima facie appeal, it would not be easy to conceive how sociobiology has come to be so attractive to such a large number of scientists. However, we have identified a number of points over which there is dispute and phenomena that do not seem to be adequately explained by this theory. More important, even if sociobiological theory were largely true, have we sufficiently understood human moral development once we know the material facts? This question and the tension generated by its opposing and conflicting answers perhaps more succinctly than any other brings us from Darwin's work and the subsequent Darwinian legacy to Nietzsche's analysis in the *Genealogy of Morals,* and its succeeding philosophical heritage.

Nietzsche's *Genealogy of Morals*: The Evolution of (A Different) Morality

No summary of Nietzsche's positions on morality can do justice to the richness of his writing (we urge readers to take up these works for themselves). The following account concentrates on the points most directly relevant to our reading of Darwin's position and the neo-Darwinian theories that have succeeded it. We then go on to discuss the broader issues that, for Nietzsche, are at stake in any discussion of the evolution of morality. It may appear at first sight that Nietzsche is closer than Darwin to a modern sociobiological account of morality. Indeed, Nietzsche's stress on the "will to power" of all life and the necessity of wholesale destruction if there is to be evolutionary advance may seem even more virulent and contrary to human decency than the calculating selfishness of sociobiology. Although these similarities seem to suggest themselves as evidence of an agreement between Nietzsche and sociobiology, as we shall see, the actual case Nietzsche makes is far more complex and subtle than anything evolutionary theory has offered us thus far.

Like Darwin, Nietzsche also contrasts our animal instincts with our social virtues. He offers us, not an evolutionary genealogy, but a physiological one that highlights cultural values in the subduing of the animal drives. Nietzsche suggests that the social and Christian virtues so admired by Darwin are not a result of any original need for cooperation, but rather are the result of coercion. They are originally the virtues of the weak, forced into submission by the strong and aristocratic; the weak were thereby prevented from freely expressing their own powerful animal drives. Nietzsche agrees with Darwin that we are miserable when prevented from expressing animal instincts, and he derives all of human moral development from this basic human condition. Nietzsche believes that this frustration of our animal drives results in a *physiological* depression. A history of the various ways humans have combated this depression gives us a history of moral development. Nietzsche proceeds to give a compelling genealogy of morals by citing evidence from a variety of areas: linguistics, Greek and medieval literature, the Church fathers, antiquated European penal codes, and Western philosophy.

Nietzsche notices that the values championed in many of the books of our own intellectual heritage are very different from those honored by the Europeans of his day. He notes, for example, that aristocratic values are unlike those of the democratic ones favored in our own time: Aristocrats, after all, are powerful and able to take what they want. To investigate and explore this phenomenon, he examines the language of aristocratic communities—the Greeks, Romans, Germanic tribes, and other Indo-European groups—for clues to the evolution of democratic values from aristocratic ones, especially examining the derivations for the word oppositions "good" and "bad" and "good" and "evil." In doing so he finds that aristocratic peoples used the word "good" ("beautiful," "blessed") to describe themselves and the word "bad" ("poor," "dirty," "unfortunate") to describe the poor, because they were the ones with the power to oppress, to demand obedience, and to name.

As Nietzsche sees it, when people oppressed by these powerful aristocrats found themselves unable to act freely on their animal instincts, these instincts did not go away; rather, they were internalized and caused the feelings of misery and frustration that are so important, though for different reasons, in Darwin's analysis as well. Nietzsche considers this internalization of instincts a decisive event:

> . . . but thus began the gravest and uncanniest ill-
> ness, from which humanity has not yet recovered,
> man's suffering of man, of himself . . . the exist-
> ence on earth of an animal soul turned against
> itself, taking sides against itself, was something so
> new, profound, unheard of, enigmatic, contradic-
> tory, and pregnant with a future that the aspect of
> the earth was essentially altered.[12]

The physiological depression produced by this forced suppression of the animal instincts has been the focus of much interpretation and its consequences fought in many diverse ways. When we feel depressed or frustrated, when we suffer, we want first and foremost to know what such suffering means. One historically powerful way of putting the facts of suffering into a meaningful context was finding out who or what is responsible for the suffering, that is, to find a cause. The first and most obvious answer was that the source of suffering for these oppressed people were their masters. This is to say that the oppressed were half right: Their oppressors were the overt cause of their sufferings, but their repressed instincts had to be present as the ground for their feelings of frustration. The downtrodden and en-slaved who had to follow their master's bidding were con-sumed by what Nietzsche calls *"ressentiment."* They wanted most to turn the tables and take their places as masters. Although they were unable to do so through force of arms, they *were* able to do so through force of words. Nietzsche calls this event the "slave revolt in morality" and claims that it was brought about principally through the agency of Judaism and Christianity.

These peoples developed their own moral standards, at odds with, but not necessarily inferior to, those of their conquerors. They reversed their oppressors' valuations; what had been virtues to their masters (power, wealth, pride, the ability to revenge) became vices. The oppressed reinterpreted their inability to subdue and conquer others (or even their own desires to do so) as virtues, and named them "humility," " patience," and "justice." According to Nietzsche, the order of naming was also reversed. Whereas the aristocrats had regarded themselves as good and in comparison saw everyone else as unfortunate, bad, or dirty, these oppressed peoples saw their oppressors and masters as *evil* and only by contrast, as a kind of afterthought, did they see themselves as the ones who must be *good*. The success of this long process, for Nietzsche, is clearly illustrated by the dominion of versions of Christian and democratic values in all the places where aristocratic values formerly predominated.

On Nietzsche's account we see that the Christian virtues that Darwin so admires come about not through any genuine softening and improvement in human nature, but through the raw power of *ressentiment* and a hypocritical interpretation of the values championed by these new masters. Moreover, this *ressentiment* still burns in the many oppressed of the world who seek to bring down the rich and fortunate and destroy their power and happiness—if not in this life, then in the next. Nietzsche interprets the lascivious pleasure shown by those who describe the horrible torments of the rich and powerful in hell as a manifestation of this seething resentment against them.

Before delving further into Nietzsche's analysis, let us look at what has been established thus far and compare it

briefly to the tenets of Darwin and sociobiology. Like both
Darwin and the sociobiologists, Nietzsche agrees that we
are essentially animals and largely motivated by instinctive
drives for power; he would, furthermore, agree with the
sociobiologists that we refuse to name these drives in ways
unflattering to us. However, the kind of mechanistic evolutionary explanation that is put forward by the sociobiologists does not really interest Nietzsche. He mentions that
the apes are a "prelude" to man, thereby recognizing a
certain relatedness, but he does not seek to explain human behavior by its underlying similarity to that of other
animals; he grants this fact but does not find it of fundamental importance. To the contrary, what interests him is
that this animal, man who is motivated by the same underlying instincts as other animals has somehow "dared more,
done more new things, braved more and challenged fate
more than all the other animals put together . . ." Nietzsche
is saying that if apes are the prelude to man, humankind,
as it is, may be a prelude to something even greater.

Nietzsche's critique of evolutionary theory does not
challenge its formal correctness. Our similarities to other
animals are obvious to Nietzsche, too obvious to leave the
discussion at that observation, or to belabor the point with
examples. He even accepts, without protest, the importance
of material factors; nonetheless, he does question their appli-
cability to all human phenomena. This questioning leads him
to wonder why the current theory of life showcases adaptation, a mere reactive principle, thereby robbing life of its
most fundamental concept *activity*. As a result, the decisive
difference between Nietzsche and modern sociobiologists is
that Nietzsche sees the possibility that we may actively form
ourselves within the constraints of our inescapable animality.
Put most straightforwardly, we can affect the system because
we are not simply at its mercy. This challenges both the radi-
cal adaptionist claims that *all* our behaviors must favor reproductive advantage and "meme" theory which allows for
nonadaptive "ideas" that selfishly parasitize our minds.

As mentioned before, Nietzsche, in some ways, seems more true to modern theories of selfishness than do the sociobiologists. They typically look for a way out of the natural conclusions of their theorizing. Nietzsche, however, is willing to accept the fact that life is fundamentally conflict, selfishness, and destruction. But for him this dark view is redeemed by the fact that life is more than a dull mechanism, it is driven by will. It is at this critical juncture that the similarities end because this is the very concept that modern science denies.

The standard scientific view is that we do not need to posit any such force as an active will to account for human conduct. Nietzsche challenges this view with what he sees as the decisive example, the one to which he devotes the entire last essay of his book, namely, the problem of asceticism and its attendant practice of celibacy, the problem that seemed so inexplicable to Darwin (as well as some modern sociobiologists). Here is how Nietzsche expresses the difficulty:

> For consider how regularly and universally the ascetic priest appears in almost every age; he belongs to no one race; he prospers everywhere; he emerges from every class of society. Nor does he propagate his mode of valuation through heredity: the opposite is the case—broadly speaking, a profound instinct rather forbids him to propagate. It must be a necessity of the first order that again and again promotes the growth and prosperity of this *life-inimical* species—it must indeed be in the *interest of life itself* that such a self-contradictory type does not die out. For an ascetic life is a self-contradiction.[13]

How are we to account for this "life-inimical species," what story can we tell to make sense of this seeming living contradiction? Here Nietzsche provides an extensive analysis of Christianity, examining it from historical, physiological, moral, judicial, and linguistic perspectives. This broad and interwoven analysis is *essential*—as odd as it may seem and as astonishing as it must appear to scientists themselves—if we want to understand the phenomenon of modern science.

Nietzsche begins by making the curious claim that one of our chief pleasures is a love of cruelty and suffering. Nietzsche marshals some historical evidence to back up his claim: Not so very long ago, no great festival was complete without a public execution. One of the chief joys of men, sanctified in their judicial codes, was to be allowed to vent themselves on some unfortunate creature freely. He points to the cruel humor of *Don Quixote*, with which we feel uncomfortable today. As mentioned before, the saintly men who authored much of the great Christian literature show a strange joy when describing the tortures of hell and imagining their enemies suffering them—if they declined from making them suffer in this life.[14] He notes the puzzling fact that being allowed to cause others to suffer has been seen as a compensation for crimes ranging from bad debts to murder. Where does this equation come from? Why is making others suffer or seeing others suffer a pleasure?

In his analysis, Nietzsche goes so far as to claim that *all* religions, at bottom, are systems of cruelty. If we are prevented from venting our essentially animal instincts for cruelty outwardly, we are forced to vent them inwardly and therefore on ourselves. Practicing cruelty on ourselves becomes a paradoxical relief from the physiological depression we feel when we are no longer able to vent these instincts outwardly. In fact, what we unknowingly seek to hurt in ourselves is precisely the animal that is the cause of our sufferings. Hence, we see the initial appeal of asceticism; it is an effort to reduce or eliminate all our animal urges, to purify ourselves from the demands of our animal instincts. Thus, we can understand the ascetic practices of self-starvation, self-torture, and celibacy, seen as the self-denial of sexual pleasure.

This ascetic ideal to which Nietzsche points us is not simple nor does it appear in a single manifestation. It shows up in many guises—the religious priest being the chief among them. The ascetic priest, through the attempted denial of his animal instincts (e.g., through celibacy, self-starvation, self-torture) is compensated in that he becomes powerful. He feels that he has power over the very ground of his being. This is a kind of subterranean road back to the freedom that had been lost through oppression by powerful humans. Even given a more Darwinian model in which humans choose to subjugate their selfish instincts to a group purpose, the ascetic gains freedom from the group norms, not by open rebellion, but by a denial of the need for such subjugation. Through this freedom from the oppression of animal instincts, ironically, the ascetic priest also acquires a moral superiority that stands as a reproach to his oppressors. Nietzsche points to the great texts of the Indo-European religions to illustrate how much power the saints of Christianity, Hinduism, and other religions have gained through their ascetic practices.

Nietzsche points out, in contradiction to the claims of most other thinkers, that it is not suffering to which we object, but meaningless suffering. If we have a reason or a goal, like the ascetic priest, we will even seek suffering. Indeed, as noted before, this self-torture may bring considerable benefits as well. The priest not only gains a certain power and freedom, but also is seen as having attained a higher knowledge with his special insight and is able to tell his flock about the reasons for their suffering and psychological misery—it is, they tell their flocks, in themselves, it is their own evil, their sin, their bad conscience. According to Nietzsche's analysis, this diagnosis is certainly false (meaning, at least, that Nietzsche does not endorse it); however, he recognized that through it the priest is able to help his flock alleviate their displeasure. Nietzsche is then able to interpret the many different guises under which Christians have lived, often with great merit, as stemming from this *belief* in their own evil.

The priest may prescribe innocent reliefs like mechanical work, living in communities (misery loves company) and doing good (small doses of power), but may also prescribe penances and the like, orgies of feeling to overcome this displeasure. In the name of contrition for sin, Christians sought, and may still seek, extraordinary sufferings that enabled them to overcome the everyday displeasure of physiological inhibitions. However, neither of these methods, the benign nor the dangerous, is able to *cure* the underlying causes which they seek to alleviate.

This ascetic ideal, this "evil eye" for all our animal instincts, has become an important aspect of our culture and is reflected in widely held values and attitudes. Since we are inescapably animals, the attempt to conquer our animality results inevitably in a devaluation of ourselves. Christianity, according to Nietzsche, has been the driving force in this devaluation.

For the purpose of contrast he draws our attention to the Greeks' attitude towards themselves as human beings and towards their gods. Typically, the Greeks had a high opinion of themselves as the central beings in the world. In the Greek epics and tragedies, when someone does something especially stupid or cruel, it is common to blame it on the influence of some god. Any number of characters have their version of the sentiment: "Humans would never act in such disgraceful ways without the gods' corrupting influence!"

Christianity, of course, has changed all this. We have gone from being the shining godlike men of the Greek tradition to being able, along with Pascal or other Church fathers, to revile ourselves for our animality—for our need to eat, digest, and reproduce. We can revel in seeing ourselves as worse than worms before the glory of God, as being utterly worthless without the support of Christ. This reflects what Nietzsche calls the "will to nothingness." Nietzsche wishes to draw our attention to the paradox of this goal, of this ascetic movement toward nothingness. Although we might be tempted to say that we can pride our modern age at having put behind us those ages in which the ascetic ideal reigned, those ages of torture, superstition, irrational belief, reliance on an ascetic priestly class who put us through hell, Nietzsche thinks we have only made a new version of the old scheme. If we have left behind the ascetic ideal in its religious sense, do we have a *new* ideal and thus a new path to put in its place? Nietzsche addresses this question:

> But they tell me it [a new ideal] is not lacking, it has not merely waged a long and successful fight against this ideal, it has already conquered this ideal in all important respects: all of modern science which, as a genuine philosophy of reality, clearly believes in itself alone, clearly possesses the courage for itself and the will to itself, and has up to now survived well enough without God, the beyond, and the virtues of denial.... The truth is precisely the opposite of what is asserted here: science today has absolutely *no* belief in itself, let alone in an ideal above it—and where it still inspires passion, love, ardor, and *suffering* at all, it is not the opposite of the ascetic ideal but rather *the latest and noblest form of it.* Does that sound strange to you?[15]

He proceeds to draw a parallel between the development of Christianity in the Western tradition and that of modern science showing the ideal is not, in fact, so new. Nietzsche now invites us to look at the progress of modern science in this light:

> Has the self-belittlement of man, his will to self-belittlement, not progressed irresistibly since Copernicus? Alas, the faith in the dignity and uniqueness of man, in his irreplaceability in the great chain of being, is a thing of the past—he has become an animal, literally and without reservation or qualification, he who was, according to his old faith, almost God.... Since Copernicus,

> man seems to have got himself on an inclined
> plane—now he is slipping faster and faster away
> from the center into—what? into nothingness? Into
> a "*penetrating* sense of his nothingness?" Very well!
> Hasn't this been the straightest route to—the *old*
> ideal? . . . all science . . . has at present the object
> of dissuading man from his former respect for
> himself, as if this had been nothing but a piece of
> bizarre conceit.[16]

It is easy to see that the trend Nietzsche noticed here—the
evolutionary view of the human species as just another animal, has been extended considerably in our time—so much
so that we can now see ourselves as merely vehicles to carry
and propagate pieces of DNA.

Yet Nietzsche did not have to wait for the develop-
ments of twentieth-century sociobiology to make his case,
he only had to look to the work and lives of the contemporary scientists of his day to support his thesis. He points to
the long working hours scientists spend, the devotion of
their life's energy to little problems. As he sees it, science is
used as a means of self-narcosis, with a view to closing off
the difficult reality of the everyday world. He is not critiqu-
ing the value or the correctness of scientific work, but pointing to the ascetic virtue entailed in the lives of scientists. He
sees further that the scientific virtues of halting before the
factual, the renunciation of falsifying, adjusting, interpretation—"all this expresses, broadly speaking, as much ascetic
virtue as any denial of sensuality. . . ."[17] Consequently, we
might say that while the world has become more and more
secular it has become no less Christian.

What underlies these two types of asceticism, the Christian and the scientific? Nietzsche identifies the underlying
commonality as a faith in Truth, a metaphysical value and
one not permitted to be questioned in science, religion, or
philosophy. Although belief in the Truth of Christian dogma
has the enlightened scientific conscience against it, Nietzsche

claims that, nevertheless, the belief in metaphysical Truth survives in its "strictest" form as the belief in science. He asserts that the belief in metaphysical Truth must be there first so that science can acquire from it "a direction, a meaning, a limit, a method, a right to exist."[18] He does not here deny the possibility of truth, but insists we need to recognize and question this underlying faith. Furthermore, irrespective of our convictions about whether or not there can be ultimate truth, the *value* of truth must henceforth be called into question as well.

Nietzsche accuses scientists of mis-valuing themselves and the meaning of their work. He claims that, far from having escaped religion and offered the world another path, they are in fact following just that path. However, this refusal of scientists to recognize their true motives and values does not constitute a reproach, or at least not only a reproach. Indeed, Nietzsche showers some very high praise on these people mixed with perhaps a hint of irony. He praises "their insistence on intellectual cleanliness; these hard, severe, abstinent, heroic spirits who constitute the honor of our age, . . . these last idealists of knowledge in whom alone the intellectual conscience dwells and is incarnate today."[19] In fact, Nietzsche credits science with getting rid of the burdensome religious trappings of the ascetic ideal and "liberating what life there is in it."[20]

What he is concerned with is not whether this or that equation produced by scientific work is objectively true, but what kind of underlying, unexamined metaphysical beliefs support the kind of effort that scientists expend. Nietzsche's questions and investigations probe deeper than checking the facts or generating scientific theory. He is concerned with why scientists do science in just the way it is done and what their most hidden desires and fears are that cause them to spend their lives in the service of science. Are these underlying metaphysical commitments true? We may not be able to answer that question. Still, these questions remain: why are they valuable to the scientist, to society, irrespective of their truth? Nietzsche wants us to see the importance of these questions and to ask them of ourselves.

The problem with which Nietzsche leaves us is this: he has argued, and this position he shares with Kant, that truth, the ideal on which science rests, is a metaphysical value, unprovable through experience or reason. Obviously, if this value is called into question, the status of science as the only road to the good and true and the status of scientists as persons devoting their life to the pursuit of pure knowledge are changed as well. Yet, as Nietzsche also points out, these men are the honor of our age, they are doing important and useful work by laboring under an ideal that is at bottom based on nothing but faith. If God or Truth are lies, we are left to wonder, along with him, if the greatest things humans have ever accomplished could have only been accomplished by sanctifying precisely lies. He raises the question not only of the existence of truth and the possibility of knowing it, but also of the value of truth regardless of this status.

In this way he brings into serious question many former notions about the connections between truth and goodness. Nietzsche questions whether the scientific results, when they challenge our most deeply held beliefs, have the power to make us relinquish these beliefs. He notes that Copernican astronomy did not really defeat the metaphysical ideal that underlay Ptolemaic astronomy—that ideal of the centrality of humanity just had to shift ground. For those who practice and believe in a scientific view—let us say the sociobiological view in which life is ultimately the mindless, mechanical replication of DNA for its own sake—science can explain our moral being and seemingly give our life a direction. But, as we have seen, others outside of science do not regulate their beliefs and lives around such scientific principles. It is not that they are incapable of understanding them, or are simply ignorant, they just hold another faith, perhaps a belief in the primacy of the mind or a creator, more dear.

When Nietzsche looks back over the history of Western civilization, he does not see convincing evidence that a reactive principle like natural selection is capable of explaining the vast array of human moral phenomena. Rather, he grants that mechanistic principles operate and instead chooses to stress that, within these natural constraints, we have somehow managed to create an animal that is distinctively human. He looks at the history of the continuing creations and re-creations of our cultures and sees the horror, the hypocrisy, the cruelty—but, and this is of decisive importance—he also sees there the capacity for, in his view, real creativity and true greatness. We have, to a considerable extent, created ourselves, for both good and evil. The rest of the animal world may have nothing so low as gulags and death camps, but they have nothing so high as Christian love either. His position seems to be that the ground of those great things we value is also inescapably the ground for the excess we most deplore. Despite these dangers, this is still a testimony to the creative power of our instinctual needs and of our ability to find ways of expressing them.

Since the Jews and Christians managed to overthrow the old value system of the aristocrats, the establishment of Judeo-Christian and democratic values has progressed inexorably. We have become tamer and more domesticated. Nietzsche agrees with Darwin on these points, though, Nietzsche's analysis goes further than Darwin's. He agrees that these virtues, so admired by Darwin, are almost completely triumphant in Europe and that we have almost succeeded in completely domesticating ourselves—but he wonders whether this is a good thing. In light of his genealogical studies, Nietzsche raises the question of the *value* of these values and explicitly challenges the traditional progressive view when he asks: ". . . the good man is assumed of greater value than the evil man, in the sense of furthering the advancement and prosperity of man in general (the future of man included). But, what if the reverse were true?"[21] Thus, Nietzsche challenges the cherished premise of the obvious preference for good over evil, just as he challenges the related premise, in his discussion of science, that truth is better than falsehood. He wonders again whether it might not be lies and twisted interpretations that have allowed us to attain new levels of civilization.

Nietzsche's Differences with Darwin

It is in the context of these larger issues that we can understand Nietzsche's comments on Darwinism and its manifestations in English philosophy and psychology. He praises the English for having made the first attempts to arrive at a history of the origins of morality. But, he wonders why they want to look in just this mechanistic direction, why they want to seek "the truly effective and directing agent" where the "intellectual pride of man would least *desire* to find it?"[22]

We might easily ask this same question of today's evolutionary psychologists. Nietzsche rejects their attempts to argue—as Darwin, Hume, and some sociobiological thinkers have—that, given the truths they have found about human morality, the only ethic we can honestly defend is one based on the principle of utility, that is the greatest good for the greatest number.

Nietzsche also disagrees with Darwin's assertion that the greatest happiness is just about the same as the greatest good and that the greatest good is the highest moral goal. Instead, as we have seen, he asserts that we aim for the greatest feeling of power, by whatever means are available. He brings to mind examples of men who have chosen to suffer, be it for political power, art, or knowledge, when other easier, presumably happier roads were available. Greatest happiness, Nietzsche suggests, may just be the rationalization of tame and mediocre men who cannot bring their deepest desires to fruition.

Is tamer better? It may be more peaceable, mean less suffering, less conflict. Even Darwin wondered and had his doubts on this issue. Another way to ask the question is: If happiness and tameness are good, good for what and good for whom? Or as Nietzsche asks "Under what conditions did man devise these value judgments good and evil? And what value do they themselves possess? Have they hitherto hindered or furthered human prosperity?"[23] We see in these Nietzschean questions the beginnings of what is usually called "social constructivist thought." He brings into question the possibility of truth, he asserts the power of words and interpretations to determine things, he accuses science of being a faith, he calls into question the possibility of any scientific objectivity. All these are common (and to many scientists, maddening) themes in the contemporary critiques of science. However, Nietzsche does not assert that scientific results are merely agreed upon by scientists, that they have no permanent status, that they could just as easily be something else. He is not really concerned with these issues at this level. In fact, he reserves some of his greatest scorn for the kind of disrespectful critic who is most annoying to serious scientists. He detests these "speculators in idealism" with their "cowardly contemplativeness" and "noble indignation." He reserves his honor for those who believe in and devote their lives honestly to science—the most powerful form in which the ascetic ideal has yet appeared on Earth. He sees that scientists are producing useful work, that their lives and work are meaningful to themselves and to society, and that science is a great enterprise worthy of the sacrifices made in its behalf. This absolute devotion to science does not, however, prove that it is the key to absolute Truth. Rather, the opposite is the case. For Nietzsche, the loud assertions of science's Truth are evidence of the lack of adequate arguments.

Let's again take stock of the three naturalistic theories of morality that we have been examining. If we compare the accounts of contemporary sociobiological writings with Nietzsche's account, we see that the modern sociobiologists assert pure meaningless mechanism, yet still want a nice compassionate society, want to somehow overcome the ruthless dictates of the genes, or memes. We might question along with Nietzsche: Whence this desire to overcome ourselves? Why would something, the description of which most resembles a robot, care in the end about a compassionate and cooperative society? Contemporary mechanistic theory is left with the difficulty that it does not seem to have a plausible mechanism to address these questions, even if we leave aside more technical issues like the evolution of celibacy.

Darwin's pluralistic approach does have a mechanism, the social instincts, which allows the Christian virtues to appear and do battle with the more selfish instincts. He argues that the origin of many of our moral traits can be explained through natural selection, but feels that other factors enable us to formulate and live by higher morals than those dictated by natural selection and the survival of the fittest. Darwin's theory therefore recognizes and attempts to account for this difficulty which contemporary sociobiology seems to want to finesse either by ignoring the problem or positing abstractions such as the concepts of memes or game theory.

Nietzsche's understanding seems closer to Darwin on some points, but closer to contemporary evolutionary theorists on others. For Nietzsche, it is possible to believe in and cultivate the Christian virtues. We are able, as Darwin also believes, to form ourselves to a considerable extent. We are not passive mechanisms. But, Nietzsche would disagree that this represents a real evolutionary change in our underlying character.

Like the sociobiologists, he claims that we misname our selfish intentions as altruistic. But, we ought not be blinded by the superficial similarity of these two views. For the sociobiologist, we have "selfish genes" or memes which are interested in their own propagation in future generations. In a sense, there are no people. For Nietzsche, there is the "will to power" which drives people to seek the maximum effectiveness or feeling of power. This may indeed result in people amassing wealth, young fertile women, or political power, but equally it may result in people living an ascetic life doctoring the poor, creating great art under difficult circumstances, or making a revolution. Taken in this view, these traits are seen as the results of an inner drive but they are not to be mistaken for the drive itself. That is to say, the results of this drive may be varied. The drive is not the cause of any particular or necessary manifestation, though it underlies all of these manifestations. The reproductive drive, which scientism posits as our ultimate drive, is here seen as only one manifestation of a deeper force. Thus, the will to power is always refracted by the preexisting conditions through which it must pass. This drive has the potential to be creative, to invent new forms of art, politics, or religion—to change the face of human civilization, but the manifestations of these inner drives are seldom creative or aimed at overcoming what we are. This is because the drives do not often break out of, but manifest preexisting structures of meaning and value.

How are we to judge, then, among these various manifestations? Nietzsche would claim that we must look to our hopes and fears about the future. It is only in the light of this perspective that we can judge where our best interests lie. The sociobiologists can only imagine a future that is essentially the same as the past, with the self same rules and goals determined by "selfish genes." Darwin seems to think that we might evolve out of that game through education or a kind of Lamarckian selection, and Nietzsche thinks so as well, though they imagine very different types of futures.

Nietzsche draws our attention to what he calls "savage nations" in many of the same ways that Darwin does, yet to different ends. He wishes to point to their values in which revenge is honored, to their severity, tyranny, stupidity, and delight in cruelty. For Nietzsche, these savages resemble nothing so much as the ancient Greeks and Romans who are so generally admired for their civilizations. All these peoples are, according to this account, essentially barbarians. They are "thin-souled" men who may be, in some ways, freer and psychologically healthier than modern civilized

Europeans. This is not, however, a recommendation to go
back to this older morality. Nietzsche does not regret the
"slave revolt" in morality that has given us Christianity. He
states that the history of humankind would have been altogether "too stupid" without this development. Yet, Nietzsche
does call into question the desirability, if we are aiming for
a higher type of humankind, of the staid mediocrity of
Darwin's dream. For Nietzsche, to achieve the highest possibilities of human life, we need to live the paradoxes that
are inherent in life, to celebrate the "unstable equilibrium
between animal and angel," not to desire their absence so
that they trouble us no more.

The point of unstable equilibrium represents the greatest danger, but also holds the greatest promise of something
better. To return to pure aristocracy or pure animality, for
all their "innocence" and "health," would be a step back,
even if it were possible. It is not that Darwin's dream of a
decent society is in itself bad; Nietzsche does not begrudge
individual humans their current contentment or happiness.
But, he is worried what this might mean for our future, he
is worried that "the present may be living at the expense of
the future" and that we, who are capable of so much, may
settle for what little we now have. Or worse. Nietzsche is not
sanguine about the peace and order of Christian democratic Europe. In fact, he feared that such regimes might
hold horrors far worse than aristocratic regimes. In this, he
was something of a prophet.

Nietzsche's and Darwin's hopes are very different. Darwin hopes people can learn to live tranquil, prudent lives
according to Christian virtues. Nietzsche hopes that we can
overcome not our instincts but our base exercise of them,
leaving behind both the old morality of master and slave and
the newer democratic morality—that we can, in a word, overcome ourselves. To do this, we need a goal towards which we
can strive. He is hopeful, though not very hopeful. His hope
lies in the fact that he sees what we have accomplished when

we have willed nothing, when our deepest desires have been toward nihilism. After all, it is in the service of the ascetic ideal that we created universal love and modern science. When we contemplate the astonishing accomplishments of humanity when they have used their powers in the service of a negative goal, we cannot help but think what we could accomplish if only we had a positive goal. The *Genealogy of Morals* is both a dark and hopeful book.

Nietzsche knows that he is making myths, but hopes they may be more excellent than other myths. Whatever our ultimate views of Nietzsche's revaluation of morality, he does point us to phenomena that have not previously been taken seriously and points to new ways in which well-known phenomena could plausibly be organized and interpreted. He at least wants to goad us into rethinking the whole question of morality from the ground up.

Nietzsche wonders why people who pride themselves on being thinkers, we educated knowledgeable persons, who have discovered so much, seem rather naive with respect to ourselves and our own motivations; why has scientism's devaluation of man been viewed as the discovery of an objective truth? What does this tell us about ourselves, and how is it that we have never thought of it this way before? Nietzsche points to a way to understand these questions. In fact, he begins his book with the following statement:

> We are unknown to ourselves, we men of knowledge—and with good reason. We have never sought ourselves—how could it happen that we should ever find ourselves? Whatever else there is in life, so-called "experiences"—which of us has sufficient earnestness for them? Or sufficient time? Present experience has, I am afraid, always found us "absent-minded" . . . we are not "men of knowledge" with respect to ourselves.[24]

We are left with at least the following question: What is the connection between the results of scientific inquiry, our practical knowledge of the world, and our knowledge and experience of ourselves? Nietzsche is thus a forerunner of twentieth-century phenomenological philosophy that attempts to deal with this question and in the process give us insight into our original experience of the world and the connection of this experience to scientific knowledge.

Chapter Four

Provoking Thought

I see brains and lips closed, tympans and temples
   unstruck,
Until that comes which has the quality to strike and
   to unclose,
Until that comes which has the quality to bring forth
   what lies slumbering forever ready in all words.

—Walt Whitman, "Vocalism"

It is not enough simply to have the facts presented to know
what those facts mean. All facts demand an interpretation
to be meaningful. Furthermore, using Darwin's and
Nietzsche's accounts of moral evolution, as we saw in the
last chapter, fundamental presuppositions or hopes for the
future determine what counts as a relevant fact, how a series
of facts are organized, and how the consequences of this
organization are articulated.

   In this chapter we explore how an objective scientific
and a more experiential philosophy can be brought to bear
fruitfully on the same phenomena.

Heidegger and the Distinction
between Ways of Thinking

Martin Heidegger, arguably the most influential twentieth-century philosopher, also spent much effort thinking about the phenomenal world and its relation to the world of science which he saw as having such a defining effect on the modern age. Though often labeled as anti-science and anti-technology, he was, in fact, immensely knowledgeable about science, especially physics, serving as a reader for scientific as well as philosophical dissertations. Heidegger was an original, sympathetic, and thoughtful commentator on modern science and its influence in the modern world as well as a close friend of Werner Heisenberg, one of the originators of quantum mechanics.

The recent English publication of the *Zollikon Seminars*, so named for the Swiss city in which they were conducted, presents us with a keen record of Heidegger's interactions with scientists. Brought together starting in 1959 by Medard Boss, M.D., medical students and scientists attended seminars conducted by Heidegger during which they engaged many of the issues we shall discuss below. Boss recounts his first encounter with Heidegger's classic *Being and Time* recalling that: "The book opened up question after question which I had never encountered before in my entire scientifically oriented education."[1] According to Boss, he was not alone in this initial reaction to Heidegger's thought. As he says of the seminar participants: "these scientifically trained doctors had never encountered Heidegger's questions as questions."[2] Quoting further from Boss will get at the heart of some of our concerns: "Many participants seemed to be shocked, *even outraged*, that such questions would be permitted in the first place."[3] Yet in the end, Boss claims his own subsequent book and the work of those colleagues and students who participated in the seminar was significantly richer for having struggled to learn from Heidegger's thought.

The first and perhaps most important aspect of Heidegger's thought that we wish to use is discussed in a number of ways and in a number of places throughout his texts.[4] For our purposes here, we shall focus on the language and way of thinking he develops in his "Memorial Address." In this piece, the primary distinction between what Heidegger calls "calculative and meditative thought" structures the essay. Heidegger argues that calculative thought is associated especially with the sciences and their particular methodologies. It is a manner of thought and a set of attendant methods that aims at correct answers or at attaining particular, well-defined practical goals. The characteristic feature of calculative thought is that the criteria for judging if an answer is correct or even thinkable are present before the investigation begins. This way of asking questions and seeking their answers stipulates in advance a set of parameters that limit the possible outcomes.

In terms of scientific investigations, Heidegger saw this kind of thought as "challenging-forth" the phenomenon in question. Challenging-forth is a dense and difficult phrase. As a broad sketch, science is said to challenge-forth the world, in part, because it sets its methods over and against the world. It shows us a world that is not *necessarily* one that can be rightly described in terms of objects and formulae, but simply one that *is* so described. It is both correct and important, in Heidegger's view, that we develop the methodologies and habits of thought to do this. When we challenge-forth things in this calculative manner, they will show themselves to us in this way, *if they can*. This suggests that there may well be significant aspects of life that forever resist objectification because they cannot disclose themselves by way of this type of investigation. However, such nonappearance is often taken by those involved in calculative thought as evidence of something's nonexistence.

Heidegger argues, especially in his essay "The Question Concerning Technology," that no matter how successful and powerful the modern scientific project may be, we should not let it completely supersede and obscure other modes of beholding or relating with the world. It is another

mode of thinking, meditative thinking, that Heidegger sees
as threatened nearly to the point of extinction by the grow-
ing influence of calculative thought. Meditative thinking deals
with questions that do not have a given correct answer that
can be stipulated in advance. Exploration in these realms
leads to what we call the "taking of a certain attitude" with
respect to the world. This is the traditional realm in which we
encounter the practices of philosophy, literature, poetry, and
music. The distinction between meditative and calculative
thinking is not for Heidegger, nor for us, a plea that philosophy, literature, and poetry be allowed some small place
on the periphery of scientific thought, or that it be a diver-
sion from or decoration for science. Quite to the contrary,
we agree with Heidegger when he says that "there are two
kinds of thinking, *each justified and needed in its own way*."[5]

Meditative thinking needs to bide its time, Heidegger
tell us. It can never, like calculative thinking, race from one
project to the next employing standard methodologies independent of situation and context. Meditative thinking allows
one to gather thoughts and to be careful: "It is in need of
even more delicate care than any other genuine craft."[6] Such
thinking cannot rest at simply reading, hearing, or noticing
things; something more is required to understand them,
namely, we need to ponder. But the delicate care, the bided
time, and the attitude necessary to ponder are all being taken
away by the overpowering, fast-paced, and computational
practices that are the trademarks of calculative thinking. Much
of great importance to us—at least as great as the project of
knowing that science pursues—can only be achieved outside
or beyond the realm of calculative thought. This outside is
the realm where meditative thinking takes root and flourishes.
Rather than "challenging-forth" the world, this meditive thinking "brings-forth" the world without demanding that it come
as an object—indeed without demanding that it come in any
particular predetermined manner.

Heidegger's view of modern science, technology, and
philosophy is instructive here seen in relief against the fore-

going discussion. Our age is scientific not merely because science is such a powerful means of obtaining knowledge or because technology is advancing at such a rapid rate, but also, and more importantly, because we have bound ourselves to calculative ideals. According to Heideggter, to be bound to calculative ideals is to be bound to a manner of thinking understood in its mode of challenging-forth. We think technologically, on this account, not because we think about technology or even because we produce technology, but rather because our thinking is in the form of challenging-forth the world as a series of objects. These ideals tell us that for which we must seek and the character of what we will find. We then feel duty-bound to comply with ideals that we have formulated ourselves and to whose power, best exemplified by the success of modern science and technological advances, we can all attest.

In order to see this clearly, Heidegger gives us a rigorous description of science from a philosophical point of view. He terms science "the theory of the real," in contrast to science's own claim to give us an increasingly comprehensive description of reality itself.[7] However, reality itself is, according to Heidegger, by definition that which cannot be captured in full by any description. On Heidegger's understanding, truth for science is that which can be "transformed into certainty of representation." Heidegger warns that "what is represented by physics is indeed nature itself—but undeniably it is only nature as the object-area . . . of physics."[8] Seeing nature in the mode of objectness, then, is only *one* way in which nature exhibits itself. As Heidegger puts it, "nature is that which cannot be gotten around."[9] Neither science nor any other way of thinking is in the position to pronounce what nature finally is. He makes an analogy with philology in which language holds sway as that which cannot be gotten around, though obviously language can be objectified as grammar, etymology, and syntax. Phenomena are always larger than descriptions of them; moreover, they will always be larger regardless of how encompassing and sophisticated the theory, either scientific or philosophical. Nature will always have a remainder and no amount of technical study can exhaust the meaning of an utterance.

This is not to disparage the pursuit of sophisticated theoretical objects. According to Heidegger, the modern university research program is ideally suited to the goals of science. The modern research experiment is a "verification of law in the framework, and at the service of, an exact plan of nature"[10] We cannot, however, simply impose such a plan and expect that nature will submit to it. This does not mean that theories are elaborated with no prior observations or knowledge, but that, in modern science, facts are things or a series of things that only become understandable in a theoretical framework. Heidegger sees the modern tendency towards increasing specialization in the sciences as a natural and salutary outcome of the fact that we must adapt our theorizing and our methodologies to the particular branch of science that we are studying. Here we might say that current books touting theories of everything are bound to miss more than they comprehend. Heidegger shows that each move toward greater scope might well be at the cost of what was best about its former explanatory power.

Heidegger's position on science is complex. He clearly sees how science, through its use of mathematics and a theory of objectness, can represent nature and set its parts in place as a "surveyable series of related causes."[11] We have created "an exact plan of nature" and the facts that scientific experimentation discovers are only understandable within this theoretical framework. He admires the way we have committed ourselves to this project and modified our attitudes and methodologies accordingly. Heidegger claims that modern science is not essentially a procedure, but that the system of science consists in a solidarity of procedure and attitude with respect to its objects. Through this formula, Heidegger puts his finger on the essence of the modern scientific view of the world: the project of objectification and representation can be unproblematically assumed to apply to all phenomena. Thus, science represents itself not simply as a particular kind of methodology, but as the only method that can produce universally replicable descriptions of the world. It is from this point of view that the objectified body of anatomy and physiology, for example, becomes the only body that we are capable of acknowledging as valid.

It is from this vantage point that Heidegger says that "Science 'sets upon' the real. It *orders* it into place so that it will exhibit itself as a surveyable series of related causes."[12] The (English) word "order" has two senses, both of which are important to the argument we are making here. Science both "orders" things in the sense of putting them into a series and into place, and it "orders" in the sense that it commands the world to show itself as a series of facts. Furthermore, when set up in this way another sense of order comes to the fore.

Once these types of scientific and calculative descriptions have been articulated, it is only a small step to the moral and political accounts of why we should "take orders" from nature so conceived. Linking this insight to an earlier example then, we are told that the objective, scientific view of ourselves as vehicles for gene replication forces us to accept the "fact" that our feelings and thoughts about our spouses are nothing other than "cold, hard equations" that our genes are evaluating with a view to their own reproduction and spread. We are told in effect to follow these dictums despite what we may believe based on our experiences of ourselves and of other ways of considering our place in the world.

Two fundamental questions are set in relief here by our explication of Heidegger: Are we, in characterizing nature in this calculative way, true to nature? and Is being true to nature the only way of being true to ourselves? Heidegger remarks that scientific representation, for its part, can never decide whether nature, through its objectness, does not rather withdraw itself than bring to appearance its hidden fullness. Science answers that it has in fact captured nature or, in its qualified version, is well on its unstoppable way—and that nature has not withdrawn, but rather has been brought out into the open and in view of anyone willing to look at the truth.

This position brings us to Heidegger's now well-known critique of science and technology. He elaborates on the question of what it means for us to order nature and what it means for us to adopt a primarily calculative way of apprehending nature at the expense of the bringing forth of meditative thinking. As we discussed previously, Heidegger had defined science as a way of ordering things and relations so they all became a surveyable series. This might be desirable for the future advance of scientific work, but this typically scientific attitude does not confine itself to this realm. It pervades, or in Heidegger's language, it "enframes" all our thinking.

Consequently, Heidegger argues, the success of science and technology harbors a danger more dangerous than any speculation of the worst adverse consequences they may cause in the world. The grave danger is that the thinking that is most conducive to producing technology has come to stand as thinking per se and become the measure by which we judge all our projects. Our age is technological to the extent that calculative thinking has become the ground for all of our thought. Technological thinking seemingly takes on a life of its own as an independent entity that needs no justification or contemplation. The result is that all of nature seems "naturally" ordered to be accessible to science's way of functioning. Technology seems to verify this. Using the descriptions of phenomena that science provides, we are able to build spaceships and microwave ovens. That they function "successfully" and as predicted seems to corroborate calculative thinking as *the* comprehensive description of nature and the only valid mode of thought.

All of nature therefore becomes something to order and to make into what Heidegger calls a "standing reserve" at our beck and call for its utility in a broad sense. But, what would it mean to say that nature has been reduced to utility? Particularly illuminating here is Heidegger's example of the Rhine river. Seen calculatively as a part of a surveyable series of facts, the Rhine is standing reserve in regard to the project of hydroelectric power, and it is also standing reserve in regard to the project of collecting tourists' money. This way of challenging-forth the river is contrasted with bring-forth the river through poetry and meditative thinking.

Another example is today's pharmaceutical companies' comportment toward the world's rain forests. From their calculative standpoint the rain forest can only be seen as a resource, as a warehouse of the future that in time will yield marketable goods. The rain forest might be left to stand, but only as a standing-reserve for future exploitation and not because it is understood as a rain forest as such. Heidegger worries that the ultimate end of this process is that human beings, who invented this manner of seeing and calculating the world, will fall prey to its demands.

Calculative thinking with its self-proclaimed monopoly on the Enlightenment, aims to make everything clear to us, so clear that we cannot help but be compelled to follow its truths. Yet this supposed paragon of clarity and freedom stands to "captivate, bewitch, and beguile" us into believing it is the only way of thinking available to us. As the numerous quotations we have cited attest, many defenders of science hold nothing but disdain for those who would question and seek paths to thinking otherwise. Against this backdrop Heidegger asks: "Is man, then, a defenseless and perplexed victim at the mercy of the irresistible superior power of technology?" This would be our fate he tells us unless we were to "pit meditative thinking decisively against mere calculative thinking."[13]

Heidegger does not deny that we can look at human beings calculatively, indeed it happens everyday in a nearly endless series of practices. He tries to get the scientifically trained doctors in the Zollikon seminars to understand: "Of course, the human being can be seen as a part of nature in the scientific manner. Yet, the question still remains whether

something human will result—something, which relates to
the human being as a human being."[14] In a wonderful attempt to show how we lose ourselves to our own calculative
inventions, Heidegger uses the example of a man face-to-
face with a blooming tree in a meadow. Science can indeed
give an account of this situation in terms of neurochemistry,
organic chemistry, and physics. However, Heidegger poses
the fundamental question for human beings:

> while science records the brain currents, what be-
> comes of the tree in bloom? What becomes of the
> meadow? What becomes of the man?—not the brain
> but of the man. . . . It will not do to admit, just for
> the scientifically unguarded moments, so to speak,
> that, naturally, we are standing face-to-face with a
> tree in bloom, only to affirm the very next moment
> as equally obvious that this view, naturally, typifies
> only the naive, because pre-scientific, comprehen-
> sion of things. For with that affirmation we
> have conceded . . . that those sciences do in fact
> decide what of the tree in bloom may or may
> not be considered valid reality. Whence do the
> sciences . . . derive the authority to pronounce such
> verdicts? Whence do the sciences derive the right
> to decide what man's place is, and to offer them-
> selves as the standard that justifies such decisions?[15]

Heidegger asks whence the standards of scientific thought gain
the authority to validate or invalidate the face-to-face experience of man. It is evidence of enframing that we are constantly
presented, through both the scientific and the popular press,
with the scientistic view of nature and humanity. When caught
by the enframing of calculative thinking, we human beings
conceal from ourselves the need for human beings. We are
exhorted to accept that all the faculties of man can and indeed
must be brought within the purview of scientific accounts. There
are huge expenditures of government and private money ear-
marked toward finding the genetic or more broadly biological

bases of consciousness, ethics, and many (especially antisocial) behaviors. It is a sign of our submission to calculative ideals that the question raised here is so seldom raised and seems to have little appeal or force in our culture. This dominion of calculative thinking does not respect the rational limits of its scientific manner of knowing.

If humans are, in many respects, a very special kind of being who, in Kant's conception, ought never be treated simply as means but always also as ends, that is as having dignity, then we need to question attempts to objectify ourselves and see ourselves as commodities whose value and functioning is merely a matter for calculation or technological manipulation. Perhaps worse is the curious conviction articulated in many scientific works that it is the height of human achievement when we conceive of ourselves as meaningless and deluded beings produced by the blind forces of the world. Once we have succeeded in reducing our total being to a calculated object of science (which, after all and despite protestations to the contrary, is the standard of sociobiology), it is clear why we are driven to make claims that deny the possibility of the will and the existence of human freedom.

We might summarize by saying that phenomenological philosophy gives us an orientation in the world that leads us toward a meditative thinking. It organizes our appraisal of situations around human intentions and projects with respect to the world. It is an account that acknowledges the human orientation toward the future and the way that this orientation organizes the meaning of space, time, and objects—humanizing them by virtue of their perceived relationships with our goals.

Merleau-Ponty and the Importance of Embodied Experience

The work of the French philosopher Maurice Merleau-Ponty attempts to understand our embodied lived experience following paths pointed to in the work of Heidegger.

In *The Phenomenology of Perception*, he gives an account that begins with our original preobjective, prescientific experience.[16] Merleau-Ponty argues that our experience of the world is primary and necessarily "before" our scientific understanding of it. Moreover, he argues that science is grounded on this experience and, somewhat ironically perhaps, much of this ground is inaccessible to the ways of knowing practiced in the sciences. He argues, then, that we need to make a careful description of this experience if we are to understand the status of the "lived world" vis à vis the world that science describes to us. In much of what follows one will recognize words and phrases drawn from everyday experience. However, they are not to be taken lightly due to their mundane origins, because in the hands of phenomenologists, they are both intended in their familiarity that is drawn from our everyday experience of the world and simultaneously raised to the level of theoretical concepts that become matter for philosophical speculation.

To demonstrate how phenomenology describes experience, imagine yourself absorbed in a favorite activity, perhaps writing a letter to a close friend, playing ice hockey, or painting. During this activity your body and mind are completely caught up in your project. Your body here is a "vehicle" for attaining your goal, your pencil or hockey stick is an extension of your body. This is the "lived body" whose "lived" space is *generated* by the body as a field of possibilities for action. In this context, space is not an objective setting, nor is space perceived as a series of measurable points. Rather, while engaged in the activity of letter writing or playing hockey, space is itself organized in light of what the body is or could be doing. With respect to these examples we can say, without absurdity, that we are near the friend to whom we are writing the letter. We might say that we are *with* our friend who may be many thousands of miles away in geographic space, just as we are *with* our hockey teammates, rather than objectively *beside* them.

Due to this immersion in what we are doing, the lived body experiences time in a manner different from the time measured in scientific studies. It makes sense to say from the standpoint of phenomenological description that experienced or lived time moves faster or slower as we are either deeply asleep, engaged in an absorbing task, waiting for an expected phone call, or watching the clock at the end of a monotonous workday. This way of giving a description of time produces accounts vastly different than those utilized in scientific measurement.

Though we all experience lived time, the overwhelming power of scientific explanation has obscured our ability to recognize and validate this time as "real." However, from those not yet socialized into abstract time we have something to learn. Children, for example, may be chided for believing an hour in the dentist's chair is longer than an hour playing with friends; however, in the context of a phenomenological examination we think that the child's honest reports are genuinely meaningful. Take your own experience as an example: What does knowing your chronological age say of the way your present actions are indebted to the whole of your past experience and your hopes and plans for the future? You carry with you in this lived moment a series of memories and possible futures that can never be understood by calculation, but only by reflection and contemplation.

Returning to a person absorbed in an activity, it is obvious that an uninvolved observer, that is, one outside the event itself, could give an objective account of these events, noting relative distances, distinct intervals of activity or rest, speed of movement, and elapsed time. This observer could give various causal accounts that rely on the mechanisms of muscle and brain operations. These accounts could even be clarified, reproduced, codified, and judged against other sets of data. However, this account would be far from the experience of the participants; it would, in fact, be an attempt to give a representation of the events observed and would do so by presupposing the idealized and objective world of scientific calculation. What escapes necessarily from these objective accounts is the very thing that most interests us here.

According to Merleau-Ponty, it is lived experience that necessarily underlies and allows for the very possibility of any subsequent scientific account. We, by necessity, always find ourselves embedded in the world, experiencing it from the point of view of our own embodiment. The objective world that science describes cannot be 'first' because initially we are *always already* a part of the world, a world we only later can come to investigate methodologically. To ask the question: "What is an adequate scientific account of the world or of a particular phenomenon?" is to have already understood the world in some more fundamental and originary manner. It is this preobjective and prescientific 'already understanding' that allows for the raising of scientific questions.

Much more can be said about experience if we follow this line of phenomenological investigation. Just as we recognize two times, the lived and the abstract, so too phenomenology shows us that we have at least two bodies.[17] We are constantly presented with the body of *Gray's Anatomy* that science describes for us. This is the body traditionally treated by Western medicine with appointments, calendars, and waiting rooms. Yet, Merleau-Ponty shows us another body, the body of lived experience. This is the body that cannot be calculated by science and of which philosophy attempts to give a description. When we are deeply involved in a project, it is the lived body and the lived world, neither any longer just as an object, that are foremost for us. However, the "objective" body and the "objective" world may intrude themselves on us and Merleau-Ponty recognizes this. One of Merleau-Ponty's most striking examples is his analysis of the phenomenon of the phantom limb (the feeling of the presence of, and even pain in, a limb that has been amputated).[18] This pain is frequently quite debilitating and it is understandably frustrating to be faced with the problem of how to alleviate pain in a part of the body which is gone. In order to highlight the contrast between the lived and the objectively analyzed body, we contrast his account with a more recent scientific account of this phenomenon given by Robert Melzack.

Melzack, a psychologist and pain expert, refutes an older model that understands the pain in phantom limbs as the result of impulses from the nerve endings of the amputated limb. He argues instead that phantom limbs originate in the brain and involve its somatosensory systems, emotional centers, and the parietal lobe (which governs our sense of self). Melzack hypothesizes that the brain produces a characteristic pattern of impulses, a "neurosignature," that generates perceptual experience even when no external inputs occur. As Melzack states: "Sensory inputs modulate the experience of the body, they do not directly cause it. Thus, it is possible to feel an arm even when no arm is present."[19]

This analysis is more complex than previous explanations and is progressive in several ways. Instead of insisting on a hierarchical structure of experience that starts only with external stimulation, Melzack's model allows for a two-way street between the senses and the brain. This alters the traditional model for pain in a limb, a one-way street, if you will, from the senses to the brain. The new model adds the dimension that the brain is hard-wired to experience a limb, even in the absence of sensory input. Notice, however, that this model, like its predecessors, treats the body objectively. Despite its novelty, it is still firmly ensconced within the calculative mode of scientific thought and is directed to the practical aim of reducing the pain and suffering in those who experience phantom limbs.

Merleau-Ponty, while not ignoring physiological or psychological explanations, attempts an analysis of the same phenomenon from the point of view of the lived body. He argues that psychological and physiological accounts both imprison us in the categories of the objective world, reducing human "being-in-the-world" to the objective, calculative world of science. According to Merleau-Ponty, being-in-the-world is a continual orientation toward the world, a continual activity that throws us into the situations of our tasks, cares, and concerns with the world.

If we begin with the realization that we always find ourselves embedded in a world that presents itself to us in a particular manner, then we get a very different analysis of the phenomenon of phantom limbs. The "world" accented in the phrase "being-in-the-world" is not the world of objects, but rather the world as it presents itself to us as a field of possibilities and meanings. This world is also an invitation to act in the context of the varying habits our bodies have developed in response to this "calling." Thus, to have a phantom limb, let's say an arm, is to continue responding to the call of the world, a call which elicits those responses for which the arm, when it was there, was primarily responsible. The body remains in the world and hears itself called to these arm actions even though the arm, the habitual medium of response to these calls, is no longer present. To have a phantom arm is to remain experientially open to all the actions for which the world calls for an arm. The presence of pain in the phantom limb is the refusal, on the part of the sufferer, to accept the contraction of his or her total being-in-the-world, despite the fact that the limb is actually missing. We might say that Melzack defines pain as an objective fact that we must work to relieve, whereas Merleau-Ponty is interested primarily in what it means to live with pain and what this pain tells us about being-in-the-world. We take Merleau-Ponty to be saying that this pain is demonstrative of the primacy of our experiential involvement with the world.

While not mechanistic itself, Merleau-Ponty's account does make use of data from physiology and medical research. Furthermore, he does not deny that there is a "way things work," nor does he deny the importance of scientific explanation. However, he approaches all explanations of this type as necessarily partial and argues that they must always be able to fit into a broader phenomenological analysis. That is to say, he never loses sight of the primacy of the involvement subjects have with the experiential world and this carries with it a conviction that no amount of scientific theory or research could fully account for this world.

Merleau-Ponty's work precedes Melzack's by thirty years. In it we find many ingredients that could point a medical researcher in the direction of a more dynamic and complex model such as that of Melzack. By thinking about the problem of phantom limbs from a phenomenological perspective that starts from our involvement with the world, Merleau-Ponty was able to gain insight into the involvement of emotions and a sense of self as components of the phantom limb experience. He anticipated by many years some of the new scientific directions Melzack takes. There are other suggestions in the phenomenological argument that could be of assistance to scientific work. For instance, Merleau-Ponty suggests particular situations and psychological states that would call up the phantom limb experience more readily than others and suggests that certain attitudes held by the subject might make a difference in whether or not one has the experience at all (not all persons with missing limbs suffer from this malady). Melzack's understanding does not take into account the differences among subjects themselves with regard to the experience of phantom limbs, because it aims primarily at relieving the pain of those who are severely affected by these experiences.

Conversely, research done in this area since Merleau-Ponty wrote corrects some errors of fact on which he based phenomenological arguments and raises new issues that demand phenomenological analysis. For instance, it turns out that very young persons with amputated limbs and those born without limbs sometimes experience this pain. Also, severing nerve endings does not, as believed in Merleau-Ponty's day, permanently stop the pain of phantom limbs. Merleau-Ponty had considerable trouble accounting for the fact that severed nerves stopped the experience of phantom limbs as the phenomenological perspective certainly would predict the contrary.

From this comparison, we are suggesting that neither the scientific nor the philosophical project is reducible to the other: A scientific explanation will never encompass our experience of being-in-the-world, and, equally important, a phenomenological explanation will never produce drugs to combat the very real and distressing pain of phantom limbs. Any claim that one of these ways is the only correct way to approach phenomena is to suggest the confusion of two worlds neither of which can rightfully claim to connect us unequivocally with the "real" world.

The sociobiologists suggest that their methods have given us just such a connection with the real. They label our experience of the world a delusion and insist that we are on the right track only when we privilege the calculative world of scientific analysis. Modern phenomenologists do not make the opposite mistake, because they do not insist that we direct ourselves solely to the world of experience. Nonetheless, the question of how exactly we are to understand and act on the type of knowledge of our bodies and the world which science can provide us is a difficult one to which we shall return. We are claiming here, using the example of phenomenology, that the mechanistic view need not (and fundamentally cannot) be the *starting* point for the consideration of phenomena. The findings of science are not irrelevant to phenomenologists but, we have seen, in the phantom limb example at least, that it is possible for the data of science to be used in a manner that transcends the mechanistic framework in which they were discovered and serve as the matter for consideration in nonmechanistic views of the world. We need not accept a simplistic either/or view in which we must either accede to scientific results together with a materialist world view or reject science altogether. This is especially the case when we are attempting to understand human thought, behavior, and culture. Philosophical thinking that respects the evidence of biology in particular and science in general informs phenomenological thinking. This is to suggest, at least, that the scientific point of view, which may be perfectly adequate as a way of *knowing* an objective world, may not be an adequate perspective from which to describe how we *understand* ourselves living in the world, and how we may live well therein.

Of course, it is possible to reject such a seemingly subjective and goal-oriented interpretation of the world as uncertain and therefore unreliable. The extreme alternative would be to accept the exclusivity of a kind of Cartesian project in which we take as relevant only that of which we can be, at least potentially, certain.

If we go back to the example of the phantom limb, we might raise the question of why medical researchers and others are interested in stopping the pain of phantom limbs. Many different sorts of answers could be given here. Of course a sociobiological response might refer to the way the desire to alleviate pain helps ourselves and our relations bear more children, or the ways in which we might expect to be reciprocated in kind by those whom we aid in this way. Another answer to why we desire to alleviate pain is that we sympathize with and are made uncomfortable by pain, both the pain we suffer and the pain evident when confronted with another who is suffering. From a phenomenological viewpoint we might say that people who truly suffer from this pain are unable to formulate projects and interact with the world in as fruitful a way as they would wish. By alleviating their pain and suffering, we may enable them to have a world again which is no longer mediated or obscured by pain and suffering.

The best use of science and its understanding of the mechanistic body is to ensure that people are not trapped unwillingly in that body, that the objective body in pain does not so overpower and oppress us that our human ability to formulate and carry through projects in the world is lost. Put succinctly, the motivation to alleviate pain is more richly articulated from our broader phenomenological understanding of it, an understanding that goes far beyond the narrow sociobiological notion of unacknowledged, blind reproductive drives. The phenomenological viewpoint is capable of acknowledging both the objective and lived bodies, while the sociobiological project ultimately acknowledges bodies only as material things.

Our everyday experiences can show us the value of this broader position. The body and mind are working correctly, from Merleau-Ponty's point of view, when they serve as vehicles for our intentions. When they cease to be these unimpeded vehicles which fall away from consciousness as we pursue our projects, then physical objects and physiological and/or psychological processes are generally highlighted instead. If we break a leg while playing ice hockey or break a pencil while writing a letter, the physical world of which we are a part asserts itself over against the intentional world of our perceptions and projects. To clarify, before beginning to write the pencil, paper, desk, even our own hand might easily be considered as purely physical objects. While absorbed in the project of writing, they are all part of a larger intention, and not considered as separate objects at all. When the pencil lead breaks, it disconnects us from our total absorption in the project of writing a letter to our friend, thus disrupting all the relations that had come together in the project of writing. The pencil that before its breaking had "disappeared" into the project now stands out from it. Our friend, who was just moments before so near to us, moves away from that nearness. Until we sharpen the lead and take up again our project, that which is participating in it (the pencil, paper, the desk, our friend) stand separated by the break. As Heidegger would have us see, the pencil is never a *thing* for us when it is being employed as part of a project.

Recall here Nietzsche's analysis of the problem of pain and suffering. Pain is easy to analyze from an evolutionary perspective. Pain evolves to inform us of when we are injured, sick, or otherwise in danger, and thus helps us react appropriately. There is obvious survival value here. Indeed, people who do not have the ability to feel pain rarely survive long enough to reproduce. Thus, we see that pain has evolved for a definite purpose and certainly still serves that purpose today. Yet having noted this, what are we to make of the many situations in which humans actually seek pain? Nietzsche uses the case of ascetics who may seek or be willing to endure quite excruciating suffering. In these cases, the meaning of pain has been reinterpreted and it has acquired a meaning at odds with its evolutionary one. In these cases pain gives a greater feeling of life and power. Its meaning has been refashioned and complicated to an enormous degree, and we see that both the original and revalued meanings can live side by side in the same individual. This is far more fascinating and should demand more of our attention and wonder than the interpretation of pain as an evolutionary trait shared by almost all animals, from the simplest to the most complex. We are not here denying the importance or validity of evolutionary explanations, but pointing to the much richer human questions that do not seem at all on the horizon of evolutionism.

Against this backdrop, we now think back to Nietzsche's example of celibacy in this same light and give an account of it from a phenomenological perspective. Recall that from a sociobiological perspective, this widespread human trait seems pointless (or at least fruitless). It has also been discounted as a relatively rare, and therefore unimportant, phenomenon that can have no positive effects on the genetic future of humankind. Scientific studies that measure reproductive success are necessarily restricted to the calculative mode of thought in which the meaning of choosing a celibate life cannot, by definition, appear at all. Thus, this rare occurrence is discounted as irrelevant, though this might also indicate that sociobiological science can only deal with the usual and ordinary and that the extraordinary is outside its purview.

Nietzsche, taking a non-calculative and more meditative stance, saw the choosing of a celibate life as extraordinary and, instead of placing it aside, for that reason made it a focus of his reflections. This self-imposed prohibition is not irrelevant, Nietzsche argues, either from the point of view of the ascetic person, or from that of people who observe or are inspired by those engaged in an ascetic life. As Nietzsche argues, such persons frequently gain an enormous feeling of power and freedom and are acknowledged as powerful by others. It is just such persons who, by way of the potency of their denial of physical pleasures, may found world religions and inspire the kind of unpredictable events in history that alter the course of the human world forever. This is hardly pointless or irrelevant, and we should not let its failure to have calculative significance inhibit our consideration of the importance of this phenomenon. From our position we are able to see celibacy as being motivated by the relations it seeks to avoid (note that these are the only relations that sociobiology is able to recognize) and the new relations that may then be able to manifest themselves. Thus, celibacy is an exemplar of how a vital human phenomenon is reduced to irrelevancy by the limited calculative view of science. Furthermore, asceticism, of which celibacy is just one form, actually re-values the meaning of pain if we grant pain the merely negative meaning attributed to it by evolutionary thinking. The ascetic actively seeks pain. Of course, this does not eliminate the negative meaning of pain as a warning of bodily peril—though it demonstrates that pain can and does also have other meanings about which evolutionary biology has little or nothing to say.

Merleau-Ponty warns that the scientific world seeks to be an explanation or rationale of the total world of our experience. We argue that scientific descriptions come after experience and we should be careful about giving these scientific descriptions primary status. If we assign a primary status to the scientific world, we at least need to give an account of why this is the correct procedure. In order to understand why scientists and others so often give a scientific world view primacy, we need to ask why this view seems so compelling as to be able to supplant our own experiences and desires as a way to organize our assessment of the world.

We can hold an informed respect for science and still dissent from the view that science is our only hope for knowledge, meaning, and salvation. By seeing that lived experience holds many important and meaningful lessons when approached phenomenologically, we mark a danger in allowing calculative thought to dominate all our thinking. We again are driven to ask, along with Nietzsche: why should we downgrade our experience and the contingency and uncertainty of the world just so we can feel safe and confident with the part of the world we are able to represent to ourselves through science? Why trade, in Nietzsche's words, "even a handful of 'certainty' for a whole carload of beautiful possibilities?"[20]

Chapter Five

The Limits of Science and the Danger of Scientism: Drawing out the Consequences for Thinking

I know the mental proclivity of physicians. I hear the chuckle of phrenologists. Theoretic kidnappers and slave-drivers, they esteem each man the victim of another, who winds him round his finger by knowing the law of his being; and, by such cheap signboards as the color of his beard or the slope of his occiput, reads the inventory of his fortunes and character. The grossest ignorance does not disgust like this impudent knowingness.

—Ralph Waldo Emerson, *"Experience"*

We have in the preceding pages reflected on the claims of scientism as to science's preeminence with respect to knowledge. The belief follows that science ought to be the guide for all decisions and actions that are ultimately of great importance to our culture. Discussions about science and society are necessarily vast and complex and there is much that has and could be said in addition to what we have attempted here. For our part, we have raised questions regarding the meaning and uses of science, particularly those based on a theory of evolution, and we have introduced an alternative philosophical perspective, using insights from phenomenology. We now extend these expositions and present a challenge to some of the oversimplified scientific positions that are especially prevalent in popular venues and that are meant to set the public understanding of what human beings "really" are.

What if we were willing to admit that the claims of sociobiology, with its emphasis on reproduction as the ultimate explanatory principle of human behavior, are valid? Even if we accept this account, though not necessarily their larger understanding of its meaning, we can raise questions that undermine the scientistic claim to be the ultimate and only legitimate knower of what it means to be human or to possess the only useful insights for therapists and lawmakers, and moreover to be the only qualified social designers of the future of the human race.

Having accepted their basic premises, where do we find ourselves? We still find ourselves in the world in which we live, with the bodies that we live in and our own personal and cultural histories, religions, and relationships with others. These are all inescapable components of the way we experience the world. We are from birth inserted into a web of meanings and from within which we learn to attach meanings to the various components of our world. Often those meanings are radically different from the ones that evolutionary scientism suggests to us. Why should we accept the "fact" that the breakup of our marriage is correctly attributed to our genes' calculations that indicate their future reproductive advantages lie elsewhere, rather than to our own search for identity and fulfillment or the breakdown of a spiritual bond that went far beyond mere reproductive benefits? In a word: Why should we decide that the scientific meaning is *the*

correct one? Scientists tell us that their work is true, fundamental, and the necessary basis for any further analysis. Consequently, we are told that we must always start with such
fundamentals that science discovers or we will go astray; we
are told it is no use living our lives as if people could be
cooperative and could build a peaceful future, because we
would be living a lie and living at odds with the discoveries
and presentations of evolutionary biology. In some instances,
as we have shown, we are simultaneously told, but given no
indication of how, we can break these "laws of nature."

Scientists are rightly criticized by many commentators
and observers for not recognizing the boundaries and limits
of their scientific work and unproblematically extending it to
the larger social world and for looking at it as a seamless
whole such that the results of scientific inquiry give us certain
knowledge of the world that can then be logically extended
to ethical prescriptions, statements about God, or social and
political agendas for a better future. Scientism's claim is that
the living world is most properly and accurately understood
through scientific investigation, that thinking about human
nature must begin and end from thinking about ourselves
qua biology and *qua* animal—because all is ultimately based
on this premise and we cannot think well unless we admit
this. Obviously, it is possible to choose this route and come
out with something that will be compelling to some people
and it certainly makes for a coherent story. We do not dispute
this. Our argument is that other ways of thinking about these
issues have always been, and should continue to be, available
as important and fundamental aspects of our thinking.

Evolutionary theory in general and sociobiology in
particular have been enormously productive ways of thinking for scientists and have led to a truly astonishing increase
in our knowledge of animal societies and behavior. Reading
Darwin's *Origin of Species* for the first time is very often a
revelation of the first order for anyone interested in understanding the natural world. This is true today and it must
have been even more so when it was first published and

evolutionary thinking was a rather new thing. Suddenly, so many phenomena of nature that had seemed puzzling or merely separate became thinkable under one grand scheme. Many things fell into place and literally thousands of new questions presented themselves to be asked and answered via further investigations of this type. We think it is fair to say that most of the research in biology since Darwin's time has been done within an evolutionary framework, or has eventually been brought within, or shown to be compatible with this framework. Since animal social behavior has been brought within the framework and looked at through the lens of evolution, we have discovered ever more richness in animal societies and know them better. Yet, it is possible to admit to a real increase in knowledge of animal communities through the use of evolutionary and sociobiological theory without also having to admit it to be the only or even the best way to understand such communities.

We have reason to be concerned with the far-reaching conclusions of sociobiological theory. There are concerns with the vocabulary used to express ideas, for instance, the use and import of words such as "selfish," "machine," and "masters" in the many contexts in which they appear. Many authors draw our attention to the disparities between the assumptions of the theory and the known facts about animal societies.[1] Although we find many contemporary critiques alluring, we do not wish to go into the intricacies of these particular difficulties here.

We are, rather, willing to grant that selfishness, in the wide contexts in which it is used, has a meaning that is something like this: actions and intentions directed toward the survival of the individual organism or the genes of the individual organism via its relatives. It is not at all a straightforward matter to know where the kind of objectifying, mathematizing analyses of science stop being fruitful ways of knowing. Surely, there is no clear line. Thus, scientists may be correct in pushing forward with their methods and manners of thinking on all fronts.

If this is so, where then do we run into problems and how do we know when the further application of scientific thinking to a problem has itself become a problem? To recognize that a theory or method of proceeding has reached a kind of limit of productiveness and utility might at least require that we be knowledgeable about what other possibilities there might be for approaching such a problem and we would also need to have our minds open to what these other lines of thinking might discover. We offer here a few examples drawn from the literature to illustrate these concerns. We do not wish to demonize these works nor, we must note, are they in any way exceptional. Nonetheless, it seems better to point to a few actual examples from among the many available to us.

Faces that Launched
a Thousand Web Site Hits

A telling example is from a paper in the journal *Nature* entitled "Facial Shape and Judgments of Female Attractiveness."[2] It has the typical tone of papers where such presumably insurmountable problems for science as the perception of female beauty are shown to yield to scientific analysis after all and the predictions of evolutionary theory are shown to be consistent with the results of said experiments. In this paper, computer-averaged faces were compared in degree of attractiveness with similar computer-generated faces with certain enhancements. In a select group of research subjects, the enhanced faces were chosen significantly more often as being more attractive. The researchers analyze these results in evolutionary terms and conclude that these data show that men are more attracted to faces that indicate a certain degree of maturity and hence of fertility. At the very least, more research is certainly necessary to substantiate this claim, as well as the claim that perception of feminine facial attractiveness has an evolutionary origin aimed at maximizing reproductive success.

These facial attractiveness studies have continued with many more publications investigating such questions as whether the judgments of attractiveness of the women in the photos are cross-cultural (it is claimed that they are) and include many other variations and permutations of the original study. You can participate in these studies by logging onto web sites and clicking on the pictures you find attractive and answering a long list of questions.[3] It is now held, for instance by Audrey C. Chen, Craig German, and Dahlia W. Zaidel that such experiments have shown us the truth behind our illusions about beauty.[4]

We are told that science has proven that "beauty is not in the eye of the beholder" in contrast to the folk wisdom to which people subscribed before research showed how wrong they were. Such research results might prompt us to ask what people mean when they assert that beauty *is* in the eye of the beholder. What is usually meant when this is asserted is, of course, that a certain couple, at least one of whom is not considered conventionally attractive, nevertheless are attracted to each other and eager to be together, to marry, and to have children. This statement is a testament to the fact that a conventional representation of physical attractiveness is far from the only criterion by which people choose their mates and that such people come to see these "unattractive" mates as truly beautiful.

The scientific results from these experiments, if their limitations are admitted, show that, on a first view, with few or no other criteria by which to judge, people are judged by their physical appearance and that certain features (such as averageness, symmetry, or height) typically influence our judgments. This is a not so exciting or controversial result, nor does it begin to exhaust the subject of human beauty.

In the tradition of the various critiques of scientism that we have so far discussed, we are able to raise the following issues: In this case, the perception of beauty is presumed to be of evolutionary origin and any result of the experiment would have been cause for speculation on exactly how this is true for the category chosen as "most attractive." Secondly, and surely more importantly, how much significance should we grant to judgments about computer-averaged images of people who do not exist and are not alive to either reproduce or fail to do so? How much does such an experiment really tell us about beauty and the perception of beauty and future reproductive consequences of this perception?

To put the question yet another way, why is such a minimal understanding of beauty worthy of our attention, research dollars, or admiration? Why should scientists be able to claim this question as solely their own, with the very explicit understanding on their part that they have finally broken another seemingly insuperable barrier to the hegemony of science to the understanding of humanity? When we think of the aestheticized uses and meanings of beauty—from Sappho and Mozart to Georgia O'Keefe and Martha Graham, or from Dante's Beatrice to Leonardo's Mona Lisa—does anyone believe that this reduction of beauty to its role in reproduction has any hope of capturing the vast and multiple meanings of beauty as they matter for our lives or count as somehow fundamental to our understanding of who we are or could become? And, if no such claim is intended, why should such results receive so much publicity including cover stories in major United States news magazines? The problem here is not that scientific accounts of female attractiveness are produced, but that other views are, on the grounds that there exists a scientific account, demoted and excluded. These scientific studies would be far more fertile if put in conversation with other ways of understanding and experiencing beauty. For this to happen, all who come to the table must be willing to put their position at risk and willing to alter opinions they presently hold. True conversation is exciting and even terrifying because of this ever present possibility.

A More Serious Concern

The work on facial attractiveness may be a rather trivial case, though beauty as a philosophical concept is no small matter in the history of Western thought as a reading of Plato's *Symposium*, for example, may well persuade. That said, perhaps a study on the evolution of rape will seem more substantial. In the scientific paper, "Rape: An Evolutionary Analysis," Randy Thornhill and Nancy W. Thornhill analyze rape as a reproductive strategy and arrive at the conclusion that it is an important evolutionary/reproductive strategy for poor men, with no other "behavioral alternatives," to get their genes into future generations.[5] Here we will leave aside any methodological problems such as data collection, sample, and research design; indeed, the authors acknowledge possible deficiencies in this regard. The authors also admit that there are other dimensions to the problem. Nonetheless, one is left, in no uncertain terms, with the primacy of the evolutionary analysis and it is hard to finesse the meaning of terms such as "only behavioral alternative." These authors acknowledge some limitations of their approach, however, they are at a loss to say what else we really should consider, how we should consider it, how other factors such as law, custom, culture, power, and the many cases where rape does not involve females of reproductive age or indeed does not involve females at all could be dealt with. And, indeed, why should they? They are evolutionary biologists and their job is to produce and interpret data within an evolutionary framework.

The authors discuss various sorts of causation in their introduction, recognizing both proximate causes (genetic, biochemical, social) and the ultimate cause of evolution by natural selection. In the end, however "the ultimate approach . . . accounts for the existence of proximate mechanisms."[6] In their conclusion the authors state that "rape is a poorly understood aspect of human behavior . . . the importance and prevalence of rape should force investigators to consider it from as many perspectives as possible and with all existing evidence regardless of how meager or conflicting."[7] They continue by asserting that rape has "defied explanation by psychiatry and the social sciences" but that their theory is "very relevant to any attempt to understand rape and any of its ramifications."[8] In fact, they see their theory as the only one that offers any real understanding of rape thus far.

We do not wish to dispute that rape can be seen as a reproductive phenomenon. On the Thornhills' own terms we might (this without disputing methodology) point out that they have not presented any data at all, such as live births resulting from rape and the survival of such children to reproductive age, which demonstrate that rape has reproductive advantages. They have shown statistically that in the data set they use females of reproductive age are more often raped. However, nonreproductive females are also raped, as are men. What this seems to indicate is that rape is, at least in part or some of the time, a reproductive act. Reproduction is one aspect of sexuality, though it is far from the only one and may not even be the most basic.[9]

Randy Thornhill published several more articles on this subject and most recently a book, *A Natural History of Rape*, with coauthor Craig Palmer.[10] This book not only repeats the arguments for rape as an evolutionary strategy, but also attacks feminist and social science positions on rape and recommends law, therapy, social policy, and educational programs based on the evolutionary standpoint. Unlike the scientific paper discussed above, the book is devoid of the usual data tables and analyses that characterize a scientific work. It is, in fact, a prime example of scientism; it is characterized by the kind of tyrannical and fundamentalist thinking that some scientists devoted to evolutionary theory try to foist on the public as truth.[11]

Thornhill's and Palmer's wholesale attack on feminists is ill-founded. This attack stems from some feminist claims that rape be understood as a crime of violence, not sexual desire. It is not clear to us why some feminists make such a sharp distinction (it seems that there can often be elements of both), but on the other hand, Thornhill and Palmer make no effort to analyze what the positive points of this position are or why some feminists are motivated to make such claims. If they had, they might have reasoned something like this: female victims of rape experience it as a crime of force and violence, as a loss of their integrity and autonomy. Invoking ultimate causes such as their loss of reproductive potential (which may or may not be a real factor) misses more than it explains. The feminist position speaks of women's desire to be considered full human beings who have a right to be free of sexual coercion and violence and their desire that courts of law recognize them as such, not as vessels of reproduction. This seems a worthy goal to strive for even if it does go against some hypothesized evolutionary model derived from speculations about our remote ancestors.

The issue of sex and violence is never addressed head on by the authors of *A Natural History of Rape*. For Thornhill and Palmer violence used in the act of rape is a means to the end sex, and for them sex can only be understood ultimately in reproductive terms. So while they admit the obvious, namely, there is in fact violence or the threat of violence in acts of rape, its presence and its consequences are occluded by their evolutionary theoretical commitments. On the other hand, many feminist positions understand that sex can be used as a means to the end violence given the complex private meanings we have attached to sexuality beyond reproduction. Consequently, from this perspective rape is a violent act that uses sex as its means trading off the meanings we have attached to each gender's relations to sex, to the wide meanings of sexual intercourse, and to sexuality as a social phenomenon. Here then we see that certain theoretical commitments to evolutionary theory rather than broaden our understanding in every instance, as is the constant promise of scientism, instead narrow it such that important elements of the phenomenon escape detailed analysis.

Another harmful consequence of a wholesale fidelity to evolutionary theory reveals itself when Thornhill and Palmer criticize a study by Peggy Reeves Sanday.[12] Sanday undertook a cross-cultural study of rape in traditional societies and based on her findings divided societies into what she calls "high rape" and "low rape" cultures. Instead of looking at this study as useful for helping us think about our society and its attitudes and helping us devise policies for reducing rape, Thornhill and Palmer attack her claim that there are even a few societies in which rape is unknown. Indeed they go to great lengths to show that there are no societies from which rape is absent. This may be true, but it strikes us as enormously relevant that there are certain societies with a low incidence of rape and that they share a group of characteristics (such as fairly equal status of men and women and a low incidence of overall violence) in contradistinction to high rape cultures.

This way of dismissing the position suggests that if one is able to quibble on one rather tangential claim—in this case the question are there in fact rape-free cultures?—then the whole of the analysis is to be dismissed. Seeking to deny one of Sanday's more tangential claims is motivated, of course, by Thornhill and Palmer's own evolutionary way of seeing the world. If rape is an evolutionary strategy evolved throughout the species, then were there a society of human beings in which rape did not occur they would be left with a theory that cannot explain rape's absence. They cannot allow their totalizing theory and calculative definition of truth to suffer even one counter-factual claim, unlike more humanist and interpretive studies.

As with other works of scientism, the authors wish to get unproblematically from facts to policy propositions. From their evolutionary perspective, Thornhill and Palmer make a number of recommendations, for example, that women should dress modestly and bring a chaperone on dates, and that boys should be given a course in the evolutionary psychology of rape before they get their driver's licenses. The first of these recommendations is not exactly news (though we also have not seen any controlled studies demonstrating that modestly dressed women are raped less frequently than "provocatively" dressed women). The second is a practice that existed long before the truth of evolutionary psychology came on the scene to recommend it. Yet it is here again that we see scientism's effects. Even though chaperoning has a long history, it is not until we have the truth from science that can we engage in this practice without illusion and with our scientific integrity intact.

Their third proposition seems odd, but is admittedly untried. Furthermore, along the lines of this third recommendation, they recommend evolutionary therapy for rape victims, teaching them about the evolutionary underpinnings of rapist psychology. They claim this would help victims understand at last (and thanks to science) the truth of their situation and inevitably lessen their trauma. This is further evidence of scientism's belief that it brings salvation–claiming that the truth of calculative thinking will set us free. Again, this therapeutic practice is untried, but we wonder why they fail to acknowledge the therapeutic strides of the much criticized feminists in creating rape crisis centers, rape hotlines, and teaching medical and law enforcement professionals how to treat rape victims with understanding and compassion.

Of course, for the authors of The Natural History of Rape, feminists are not the only ones who have misunderstood rape from being insufficiently evolutionary in their thinking. Thornhill and Palmer, along with many thinkers in their fields, are also fond of criticizing Freud and other psychologists who theorized about men and women's sexual desires and the deep, complex, and variegated role sexuality plays in our psyches. They dismiss without analysis the entire Oedipal dynamic as well as the claim that some women express in fantasy the desire to be raped with the observation that "in reality, any desire to be raped must always have been selected against in human evolutionary history, since it would have interfered with the fundamental reproductive strategy of females—that is, to choose mates on the basis of the benefits they are likely to provide."[13] This response begs the question at hand in the guise of the definitiveness of evolutionary truth.

It is certainly the case that these are matters for thoughtful discussion, but it is likewise the case that, for example, Freudian claims are reflected in much of the world's literature. For instance, consider those reports where women's fantasies of rape are a part of their sexual lives. To have such sexual fantasies is not, of course, to say that women ask to be raped. To the contrary, it shows the complicated nature and vicissitudes of our sexual lives and our desires. Furthermore, even the sexual desires at play when two persons make a conscious decision to procreate are likely to harbor nonprocreative concerns as well. The cavalier dismissal of any psychological mechanism not consistent with evolutionary psychology, despite the fact that people really do experience and embody thoughts such as these, is scientism pure and simple. Freudian analysis has profoundly expanded our understanding of the human condition and the literature dealing with it has given us much to consider and to contemplate; in addition, it allows many insights into personal, family, and societal dynamics. This is reason enough to keep Freud and other nonevolutionary psychologies on our list of positions to be considered in discussions of sexuality and rape.

Freudian analysis is not alone in this regard. Briefly, an analysis indebted to Karl Marx, for example, would have much to say about these issues as well. The finding that men of low socioeconomic status predominate in the statistics on rapists may be an indication, as the authors of the original paper suggest, that they have been reduced to their "only behavioral alternative." However, it is a Marxist-inspired analysis that could show how such social categories are constructed in the first place. Such an analysis would have quite different readings of the variables used in this study. Power, wealth, relation of bosses to workers, to name a few, would all lead to an alternative and we think credible story about the meaning of rape in a class-based society. It would also help explain why higher status men, though also engaging in rape, avoid appearing in the statistics.

In the end, even if rape were fundamentally a reproductive phenomenon, how can this realization help us with its personal, social, and legal ramifications? Do we punish rapists less harshly because it is in their nature? Do we punish poor rapists less harshly than rich ones because rich men have easier access to nonviolent intercourse with women? Do we punish poor rapists severely because such examples are the only ways to make the "cost/benefit analyses" of other potential rapists tilt over to the side where the cost of the attempt seems too great? We cannot find the standard of judgment for these questions from within science. It must come from an independent ethic. Here, for instance, Kant would say that it is immoral for one person to use another simply as a means to his own pleasure or profit. On this ground, rape is simply forbidden as a categorically immoral act.

What social science, political economy, and psychoanalysis offer us at their best are powerful diagnostic hermeneutics, each giving us a way to see in our social world phenomena that would have otherwise remained hidden from us or little understood. All these hermeneutic strategies give us a wealth of information and offer considerable explanatory power concerning human behavior and social interaction. This information has been produced by decades of studies and intellectual work taking as their objects the human social world and the human psyche. Should not we assume that the insight derived from these decades of work is at least as relevant to our understanding of the richness, multiplicity, and historically intricate human social world as the speculations derived from evolutionary psychology?

We remain convinced that the human social world is too vast, complex, and multiplicitous for any reductive theory to claim all of our intellectual attention. Everything is not the result of political economy, though it would be unwise to ignore the role it can always be said to play; everything is not the result of the psychosexual dynamics of the id and unconscious, though it would be unwise to ignore the role they can always be said to play; and finally everything is not the result of reproductive strategies, though it would be unwise to ignore the role they can always be said to play. We can see from considering even this brief analysis of the psychological, social, biological, and legal ramifications of beauty and rape that it can never appear as a pure phenomenon and that we are arrogant if we think that these or any other human action can be ultimately reduced to or understood by an evolutionary account.

Discussions and deliberations concerning the rational and just organization of human action need to take place in the public sphere where arguments and evidence can be proffered and addressed. This public sphere has to be free of tyrannies either enforced physically and/or ideologically. Experience suggests the larger sociopolitical world of argumentation, reflection, and deliberation needs certain guidelines. To conclude our investigation, we shall think through some of these requirements and attempt to show how in this all important arena we might find ways to live well taking off from the other side of our Enlightenment and cosmopolitan heritage.

Chapter Six

Leaving Us to Wonder

In a footnote to *Perpetual Peace* Kant makes the suggestion that one should rather presuppose that nature and freedom work towards the same end than flatter tyrants and abuse the human race.

—Agnus Heller, *Can Modernity Survive?*

The Difficulty Begins Here

It requires neither research scientists nor philosophers to persuade us that much is wrong and much is missing in the world today and that the problems are serious and pressing. However, after we agree with this rather obvious point, where and how do we proceed? Reflecting on this state of affairs, the French philosopher Pierre Hadot asks a question that can perhaps send us on our way:

Isn't there an urgent need to rediscover the ancient notion of the "philosopher"—that *living, choosing* philosopher without whom the notion of philosopher has no meaning? Why not define the

117

philosopher not as a professor or a writer, who develops a philosophical discourse, but, in accordance with the concept which was constant in antiquity, as a person who leads a philosophical life? Shouldn't we revise the habitual use of the word "philosopher" (which usually refers to the theoretician) so that it applies to the person who practices philosophy, just as Christians can practice Christianity without being theorists or theologians? Do we ourselves have to construct a philosophical system before we can live philosophically? This does not mean, of course, that we needn't reflect upon our own experience, as well as that of philosophers of both past and present."[1]

Despite the fact that the dialogues that record Socrates' philosophy, which he accomplished through living and speaking with others, were written well before the cut off date for relevance of 1859 (advocated as we saw earlier by a number of scientists), we find good reason to return to them and not simply because they may have "considerable" historical value.[2]

The philosophy of Socrates takes seriously the root meaning of the concept philosophy; it is the "love of wisdom." Socrates' philosophy studies life as dynamic training for living well or, as Hadot puts it, we ought to understand philosophy as exercises in developing a sound and ethical "way of life." Along with us, Hadot believes Socrates an exemplar of the practice of philosophy as a way of life and thinks Socrates stands today as a model still to be recommended. The love of wisdom pursued by Socrates is not a system of ideas primarily, but a way of life as an orientation toward living and as such a never-ending quest. Such a quest for a way of life has produced many paths that, although they may well cross, are not required to coincide. All these philosophical paths are based on a shared belief that we must search and reflect daily, that the way is never secured once and for all. In a word: Philosophy *gives life over to those who have to live it* in the struggles of everyday existence and the concerns they have for the world and everything in it.

Against this backdrop we can see the stakes and differences between the ways we are advocating for pursuing questions about life and living well and those advocated by scientism. This is not, however, a choice between two philosophies. Scientism is not philosophy foremost because it does not love wisdom as is the first priority of those who are philosophical, rather it is the love of facts and theories. A commitment to search continuously is one that would never think of obtaining a definitive "theory of everything" in the sense advocated by today's scientism. The move toward knowledge and enlightenment as embodied by science is unfortunately marred by a countermove to authoritarian thinking that goes against that spirit of enlightenment whose torch scientism claims to carry. Given scientism's blanket dismissal of other approaches to complex human problems, the claim of calculative thinking's superiority to all other intellectual approaches, and its belief that all problems should be solved by following their dictates leads to a situation in public life that people are right to reject as misleading, if not in fact dangerous to democracy.

Let us look at some of the reasons we should not look to scientism or a reductive scientific world view for guidance in living well. Historically, powerful regimes have used the prestige of science to bolster and legitimate political and social practices. In *The Origins of Totalitarianism,* Hannah Arendt examines the practices of National Socialism, a regime that explicitly stated that "politics is applied biology." Their theorizing went far beyond the now oft discussed identification of innately inferior and superior human races. The official view of humans as mere matter and function, who could be completely controlled by reducing physical life to its most basic instincts, guided theory and practice in the European concentration camps in the middle of the last century. Social Darwinism (the broad application of the theory of "survival of the fittest" among human social groups) has come under fire from contemporary evolutionary thinkers

for precisely this reason—because of the heinous practices
of the National Socialist and other totalitarian regimes, which
used theories ostensibly drawn from scientifically confirmed
evolutionary principles.

Thinking Beyond and Otherwise

Aware of the evil done in its name, "New Darwinians" such
as the evolutionary psychologists assert that they do not fall
under the critique of this older, more totalitarian social
Darwinism. Using Darwinian principles, they advocate kinder,
gentler values. However, as we have argued from the start,
it is naive to think that any conclusions drawn from evolutionary premises could stand as the only possible ones. From
experience we know that Darwinism can serve scientists and
others as material from which to draw both conservative
and progressive social and political implications. Arendt
asserts, for example, even under the most materially reduced
of conditions, human spontaneity of action continually
manifests itself. This manifestation she calls "miracle" and
this understanding leads her to maintain:

> Hence, it is not in the least superstitious, it is even
> a counsel of realism, to look for the unforesee-
> able and unpredictable, to be prepared for and to
> expect "miracles" in the political realm. And the
> more heavily the scales are weighted in favor of
> disaster, the more miraculous will the deed done
> in freedom appear; for it is disaster, not salvation,
> which always happens automatically and therefore
> always must appear to be irresistible.[3]

She continues along these lines arguing:

> Objectively, that is, seen from the outside and
> without taking into account that man is a begin-
> ning and a beginner, the chances that tomorrow

will be like yesterday are always overwhelming. Not quite so overwhelming, to be sure, but very nearly so as the chances were that *no* earth would ever rise out of cosmic occurrences, that *no* life would develop out of inorganic processes, and that *no* man would emerge out of the evolution of animal life. The decisive distance between the "infinite improbabilities" on which the reality of our earthly life rests and the miraculous character inherent in those events which establish historical reality is that, in the realm of human affairs, we know the author of the "miracles." It is men who perform them—men who because they have received the twofold gift of freedom and action can establish a reality of their own.[4]

Experience is a living category that escapes all attempts of scientific reduction and can never be completely captured by its system of fact production. Experience, which is always beyond our ability to predict and control it with certainty, shows us the fundamental character of our being together and our embeddedness within communication and community. It is this always already being a part of an ongoing communication that leads us to turn our attention to the public sphere as the site of dialogue, and ideally the place where we start the search for a well-lived life and where, we wish to argue, democratic decisions ought to be made.

Our concern for democracy is best seen as a concern for the public good, how we think about it and how we can help to accomplish it. Scientists often position themselves rhetorically on the side of the public good. Science, on a practical level, can help ease pain, help feed the hungry, and help ameliorate any number of social ills. This, along with its claims to know the truth, gives science considerable power and influence and helps deflect criticism from those without scientific credentials. On the other hand, in the same way that scientific theories often served powerful and unethical interests in totalitarian regimes, in today's discourses the prestige of science is often aligned with military and corporate powers and frequently serve ends that are arguably far from the public good.

Turning to the Public Sphere

As a way of going on to a discussion of the social and political implications of our line of thought, we take another look at one of Kant's questions which we discussed earlier: For what may we hope? This question implies an orientation toward the future and what we can make it. If we believe only in destiny or genes, we can only hope for a future that is like the present and which we have little or no power to affect in significant ways. If we believe that there is some freedom, we can at least be co-creators in the world. This does not mean that we can make the world whatever we care to envision, but it does mean that we can conceive of ourselves as a cause among other causes and it may mean that our visions, imaginations, and fictions though not realizable in any absolute sense, still matter and may well matter most of all.[5]

Thinking about possible futures and the problem of human freedom is not just a difficulty for thought and every individual's search for meaning. These views are disseminated in both academic and popular writings and lectures and do have real social and political consequences. Indeed, at some level, they are certainly meant to do this, for some authors more overtly than for others. These battles over what constitutes truth and where our highest duties ought to lie filter into the world of our day-to-day relations with other citizens and may profoundly affect those relations and thus affect the kind of decisions we can come to as a political community. Of course, they are just one aspect influencing the political directions a community may take, but their rhetorical value in political debate is frequently very strong.[6]

Kant argues that freedom of thought is necessary for a world in which there is hope for a better future and that such freedom is impossible without dialogue and conversation. Opposition to our freedom of thought is threatened first by civil coercion and secondly by what he calls "moral constraint." Of the former he writes:

> We do admittedly say that, whereas a higher authority may deprive us of freedom of *speech* and *writing*, it cannot deprive us of freedom of *thought*. But how much and how accurately would we *think* if we did not think, so to speak, in community with others to who we *communicate* our thoughts and who communicate their thoughts to us![7]

Kant goes on to argue that without dialogue with others in a public space fitting to the endeavor, our thoughts come to nothing and our freedom is removed from us. As such, freedom of thought in this depreciated condition is only a freedom so-called because it has no force or consequence. A full-bodied freedom of thought needs others first and foremost as without others there is no possibility for the public and democratic exercise of reasoning.

Yet this is not the only difficulty for the practice of public dialogue. The second coercion about which Kant writes is a religious devotion that is too zealous and thus authoritarian. Scientism as a new religion can be submitted to the same critique that Kant puts forth here against any religion that oversteps its boundaries. Kant says this constraint comes about:

> whereby some citizens, without the use of external force, set themselves up as the guardians of others in religious matters, and succeed in outlawing all rational inquiry—not by argument, but by prescribing articles of faith backed up by a nervous fear of the *dangers of independent investigation*,

impressing these articles from an early stage on
the minds of those concerned.[8]

Certain defenders and proponents of scientism come to us
with their unquestioned claims to know truth and a complete reliance on the utility of calculative thinking. The
dangers to today's public sphere are thus placed in stark
relief. Kant's concern for religious authority, of which we
are right to take keen notice today, is apropos, mutatis
mutandis, of scientism as well.[9] Perhaps it is worse with
scientism today as it receives untold amounts of public funds
and research dollars to pursue its goals and its efforts are
supported by many corporations as well as government agen-
cies. Be that as it may, the most dangerous aspect may well
be that scientism presents itself as the great *secular* savior, a
move which serves to hide the structure of its thinking and
speaking, and one it shares with certain other religious discourses. We cannot insist on the importance of this enough:
Neither a religious theocracy nor a scientistic technocracy,
both of which are believed to be based on an unquestion-
able foundation of authority and power, are fitting social
organizations for free and reasoning beings.

Consequently, if reasoning about the public good is to
be open to us as a democratic public, then we need others,
many others, originating from a variety of perspectives to be
in dialogue and conversation with one another. Furthermore,
we cannot allow any one manner of approaching the question at hand to tyrannize other perspectives. Though one
perspective may well come to win the day, as it were, it
cannot do so because of any guarantees it is granted from
the start, but only through the persuasiveness of its arguments and evidence as they appear in public and as they
provide thoughtful answers to their critics.

The view we articulate here is neither a synthesis nor
a middle path, but accepts the constant influence of dispar-
ate intellectual and experiential forces. It accepts the ability
of science to give fascinating and beautiful explanations of

many aspects of the material world, it sees the potential utility of science in many ways while it recognizes, as well, both its seductions and its limitations. What is more, we do not want to put ourselves in the position of inverting the scientific view and asserting the subjective is "true and real" and that the sign of truth is not the objective calculations of science, but our deepest feelings of the meaningfulness of life, be they religious experiences, the feeling of communion with nature, or any of a myriad of other possible forms. This opposite extreme is just as disingenuous as the first path of scientism. The pure subjective experience, no matter how powerful, is also not indisputable evidence of absolute truth. The mere fact that such experiences are so different for different individuals, or for the same individual at different times of life, should be evidence enough of this. This lack of absolute truth emphatically does not reduce the *value* of these experiences, though it certainly may complicate the matter when it comes to reaching agreement among many different sorts of people as to what is most important. Yet, as we have gone to some lengths to make clear, to admit this difficulty is not to admit that we should favor the claims of science to decide these questions. It is, and remains, a difficulty with which we must struggle. We are better off to prefer the existential struggle of living well to any *easy* answers that make the problem appear to disappear.

As we do not seek nor promote an easy path, this study is not undertaken to end with a series of infallible prescriptions for ethical behavior. Indeed, as should be clear by now, we do not believe any such formulae exist. In the main, we see what we have accomplished to be the raising of questions that are rarely if ever posed by those writing about science, particularly in public contexts. We are concerned with the totalizing thinking that often characterizes the writings of scientists in works intended for the general public. We are also concerned that persons who are intuitively suspicious of such totalizing thinking be presented with alternative ways of interpreting situations. Thus, our final move is to think through some issues of the public sphere. Note well that this is not offered as a panacea to all the problems we have tried to cause for thinking. It would take another book-length study, at least, to begin to deal fully with the issues we have raised in respect to the public sphere given, for example, its current domination by corporate media owners and their corporate advertisers. As we do not think that we can aspire to ultimate solutions, we do not, therefore, suggest that our concluding discussion is an attempt at closure, but rather an invitation to dialogic struggle.

The being-together we share in public life is today almost entirely mediated by a technocratic approach to problems. It follows from this that the values we most often see at work in our social and personal lives are indebted to the assessments of calculative thinking. The whole of our lives is, by and large, determined by these technocratic values from many of the everyday encounters we have with other human beings, to national foreign policy.

The Czech dramatist and political thinker who became Czech president following the 1989 Velvet Revolution, Vaclav Havel, is another important ethical and political thinker whose way of engaging these pressing questions can give us some requisite guidance. Spending many years as a prisoner of conscience due to his dissidence during the days of the Soviet-sponsored repression in then-Czechoslovakia, he found ways to produce writings critical of the calculated reduction of human freedom and responsibility. While in prison, he wrote essays and plays exploring the roots of totalitarianism and considered how we could best respond to the threats these systems pose. In "Politics and Conscience," Havel expresses alarm at the rationalist spirit that animates much of contemporary politics, including science and modern life in general arguing there:

> So, too, the totalitarian systems warn of something
> far more serious than Western rationalism is willing
> to admit. They are most of all, a convex mirror of

the inevitable consequences of rationalism, a gro-
tesquely magnified image of its own deep tendencies, an extremist offshoot of its own development
and an ominous product of its own expansion.[10]

Havel unashamedly draws on his experience to argue that
such ultrarational world views are devoid of any human
reason but, are instead "a monstrosity which is not guided
by humans but which, on the contrary, drags all persons
along with its 'objective' self-momentum . . .' "[11]

These ultrarational systems are dangerous because they
are devoid of human beings who are willing to take responsibility for the consequences of their thoughts, speech, and
actions. In order to assume this kind of responsibility, we
must speak and act from a place in our world and from our
experiences of that world, rather than from some theoretical viewpoint which is presented as unassailable, because
rational and impersonal. Of course, this does not mean
eliminating science, technology, or politics but "rehabilitat-
ing the personal experience of human beings as the initial
measure of things . . . in returning content to human speaking, in reconstituting, as the focus of social action, the au-
tonomous, integral, and dignified human."[12]

The genuine possibility of engaging in the public sphere
requires that before we enter we have the habits of freedom
and thinking as an integral part of our experience. However,
in a world given over to calculation, we seldom are given the
opportunity to exercise free and thoughtful deliberation.
Without this difficult and demanding practice, we too often
find ourselves unfit for the public discussions demanded of
persons assuming the responsibilities of citizens. Furthermore,
the vast majority of political issues in the terms and manners
in which they are presented to us, have already been framed
in technocratic ways and thus disguise from us our need to
investigate them by some other nonmechanistic means. That
is to say, when presented with an issue in calculative terms it
invites if not demands from us a calculative reaction.

This need for the habits of freedom and responsibility are also addressed by Havel, this time after he had become president of the Czech Republic. After the years of political domination as described in his earlier essay, Havel finds that granting political freedom to those unaccustomed to it does not necessarily make for a smooth transition to democracy. Without habits and the practice of freedom, the first consequences of the new freedom lead to much behavior Havel saw as "the dazzling explosion of every human vice." However, rather than attributing this consequence to selfish genes or an unalterable human nature and thus an argument against freedom, Havel understands it as the need for a "complex, long-term, and never-ending work involving education and self-education." Thus, the question arises: what, then, shall we teach if we want a politics of morality and civility? What do we need?

> All you need is tact, the proper instincts, and good taste. One surprising experience of "high politics" is this: I have discovered that good taste is more useful here than a post-graduate degree in political science. It is largely a form of knowing how long to speak, when to begin and when to finish; how to say something politely that your opposite number may not want to hear; . . . how to keep the conversation going without being aloof. . . . Science, technology, expertise, and so-called professionalism are not enough. Something more is necessary. For the sake of simplicity it might be called spirit. Or feeling. Or conscience.[13]

The fate of democratic freedom in the United States today is more precarious, perhaps, than the situation Havel describes from two vantage points, in the United States we have not the overt forms of oppression so easily recognized in Havel's situation and we already perhaps believe we have good taste and are well mannered enough. Despite this, or possibly just because of this, we must remain ever vigilant in the face of the questions: Have we the genuine habits of freedom today and do the often hidden and undetected consequences of scientism threaten what few good democratic habits we might be said to embody? The seemingly smooth running of much of everyday life might well seduce us into thinking, wrongly we suggest, that we have more than enough freedom, more than enough good taste, and more than enough democracy.

Returning here to the thought of Arendt in her beautiful essay "On Humanity in Dark Times"[14] helps us understand what is at stake in these and many attendant questions. In this essay, reflecting on the example of the writer Gothold Lessing, she wonders why so many people feel that their highest duty is to objectivity rather than to developing the habits of freedom and friendship (read democracy). She ponders about the "correctness" that science seems to offer today and about those other historical times when God's truth filled this conceptual place. Both, she argues, seek to reduce our understanding of the world to one set of unassailable facts. In light of this she wonders why could we not think of friendship and never-ending, open-ended discourse with others in our community as our highest duty rather than sacrifice friendship and freedom to objectivity?

Perhaps a discussion of a recent work in evolutionary biology connected to a pressing social issue will help to demonstrate how this insight continues to have relevance for thinking today. A widely discussed thesis by evolutionary biologist Joseph Graves appears in his *The Emperor's New Clothes*.[15] Graves claims "the purpose of this book was to remove the scientific foundations of claims that genetic or biological differences in human populations determine their social conditions."[16] In this work Graves claims to prove scientifically that there is no biological basis for thinking there are different races of modern human beings. Furthermore, Graves claims for his study a place in the larger sociopolitical struggle against racism. He is well aware that pernicious practices of racism exist in America and goes to great lengths to make it clear that from a scientific perspective there are no human races is not to say that there are not behaviors based on using a common sense understanding of race as the salient interpretive factor. Nonetheless, his argument is that all such common sense understandings are wrong because they are scientifically false, and all uses of the concept of race in these instances are harmful illusions. He is hopeful that his study will counter these damaging falsehoods because he believes that they are all based on demonstrably "bad science." In the end, Graves claims for his study the merit of following scientific truth on this score, concluding with an air of scientistic confidence: "thus one does not have to compromise one's scientific integrity to be morally opposed to racism and other forms of bigotry."[17]

Indeed we agree that overcoming the insidious consequences of racism is a pressing human goal. However, we note that Graves qua biologist struggles on rather dubious grounds. How can biologists committed to the methods that search only for the truth know what they are going to find before they begin their studies? What if Graves had found that there were in fact much statistical data scientifically suitable for distinguishing biological human races? How could we be certain that we would not find such data before we began? What if new data comes to light, based on the very methods favored by evolutionary theorists, and we find that there are in fact nontrivial hereditary differences among races in humans and thus clear distinctions among them? What if this data proves that the race historically called "white" turns out to be inferior to all other races or any such finding potentially detrimental to one group's well-being? Where do we turn then for our moral guidance—not to biology or science it would seem; would this be the time to "compromise our scientific integrity?"

Ought we have followed this argumentative strategy in the first place? The thinking that will be necessary for us to

be ethical and to have a chance at being just must come
from somewhere other than our consideration of scientific
theories and facts. We must not ask how we should live our
lives given a set of facts and theories, but how such facts and
theories can contribute to living well. Our political lives
cannot be run solely by the facts of biology (or any science)
and not just because it can be bad science but because it
might someday be good science that is used for discrimina-
tory and unjust purposes. Though popular across many
disciplines, perhaps Stephen J. Gould's famous *The Mismeasure
of Man* ought to undergo a similar critique. Indeed we might
say, in the end, that man was missed *because* he was measured not simply that he was mismeasured. We suggest
that thinking living well starts neither from materialist
science nor a scientific world view, but from our original
embeddedness in a communal world and our experience of
that world.

Arendt's thinking, influenced by having lived through
the dangers of the rule of a totalitarian regime, gives us an
outline of some of the philosophical considerations we ought
to address in our attempts to participate as both intellectuals and citizens intent on living well in a democratic community. As a student of Martin Heidegger and as one who
well understood the philosophical concepts of which we have
made use in this text, Arendt wrote insightfully about the
danger of a world reduced to calculation. Her philosophical
thinking confronts in ways akin to our questioning the
implications of calculative thinking for democracy.[18]

Thinking (as Arendt understands the concept of meditative thinking from Heidegger) is an inherently unproduc-
tive activity. This means that unlike calculative thought it
does not—and is not meant to—lead to the production of
utilitarian objects of any kind. Nor is it meant to lead to
definite knowledge or certain conclusions. At first blush, it
may seem useless for us to spend our valuable time contem-
plating the problem of human freedom in all its difficulties
and subtleties given that we can never be sure, and when we

seem to have satisfied ourselves as to the correct conception one day, the next day we must start again and may even reverse our position. Yet to participate in philosophical exercises as a way of life is to realize that such seemingly useless enterprises could neither be more important nor could they recommend themselves too highly.

Such philosophical exercises as the habit of taking up, turning over, and meditating upon issues that come to us through our experience—considering them for their own sake—is, Arendt argues, a habit we need to cultivate for our own ethical well-being and as an aid to the social and political well-being of our communities. By cultivating the habit of taking up and examining such claims with regard to premises, logical consistency, broad implications, soundness of thought, and relation to other thinking on the same topics, we can steel ourselves against the typical strategies authoritarian regimes use to control their citizens.

Although this type of exercise is, from the point of view of practical concerns, useless in that it produces nothing tangible, it is not without consequences because it is linked to another of our intellectual faculties, namely, the faculty of judgment. Without the practice of reflection and a type of practically useless meditative thinking, our faculty of judgment would atrophy as with any unexercised part of ourselves. The ability to judge well the questions of the common good requires a strong faculty capable of undertaking the task and this faculty is made strong enough for the task at hand by the practice of thinking. The mere application of some calculative equation, however complicated and precise it might be, is not on Arendt's account enough to allow us to say we have made a judgment. Judgments require facts, perhaps many of them, but they occur always some way beyond those facts. There are many situations in which we rightly feel we have too few facts to judge well, but we shall never be in a situation where the facts themselves alleviate our responsibility for having to make a judgment. Making a judgment always requires a human being who in her judging moves beyond the facts as such and links them in some way with the living concerns of those involved. This linking of facts to human life requires something other than calculation and just the mere presentation of the facts themselves.[19]

Desire for Public Dialogue

People increasingly despair of participating in the public sphere, feeling much more comfortable when they can withdraw into their private brotherhoods and live among people who all think and believe just as they do. The problem with this, Arendt argues, is that the truths professed in their small brotherhoods become the only truths that they can recognize and they then profess them to be the only possible truths.

In this way the public realm, which only operates fruitfully when we are willing to consider perspectives other than our own, is rendered impotent and thus people lose the very skills that they need to operate effectively with one another as interlocutors. The diminished public sphere that results is a danger to us all. As almost everyone now laments the realm of public life is now left open for those with the most violent and illiberal views, with simplistic and dogmatic positions, and with the thinnest slogans and sound bytes. This diminished way of conducting public life is frequently the cause of great harm and further divisiveness in our communities and drives us ever further from productive participation in the public realm. In this way, the configuration of money, power, advertising copy, and all that these taken together imply, will decide even the most important public questions rather than deliberative judgment in the sense we have outlined above.

The past century has seen numerous regimes that have made use of the language of biological determinism with its concurrent doctrine of the absence of human freedom, of humans as mere matter and function. Acting on this belief and inculcating others with it has been the ground in which totalitarian regimes have prospered. The religious-like fervor under which atrocities are committed should be a further warning to us about the dangers not of God or science per se, but of the single-minded and limited world that totalizing views may foster. Although it is far too simple to ascribe the rise and success of totalitarian regimes solely to a belief in biological determinism or the absolute effectiveness of scientific explanation, Havel's and Arendt's arguments allow us to see it would be naive to think that such views, especially when widely disseminated and accepted in the public realm, have no consequences. It seems a particularly ominous situation when the public has become accustomed to, and consequently demands, simple calculative explanations for complex issues and when it is always ready to embrace the next simplistic view that requires no further thought or puzzlement on its part.[20]

It is not adequate merely to recognize the dangers of fanaticism and eschew them. The fervor of a belief in science or some other principle can and does lead to new discoveries about the world or the human soul that could not ever be accomplished with a calm, detached armchair approach to the world. Such great faith is ground, as philosophers from Plato to Nietzsche have long contended, for both great good and great danger. So, the question may be best phrased: If we want the intellectual and practical benefits that pursuing such lines of thought can produce, how do we also guard against the worst consequences of this freedom? To repeat an earlier point, we do not wish to argue that the prudent path would be the middle path of mediocre faith and mediocre results. Rather, we hope that by cultivating the practice of seeing the world through a variety of hermeneutics and by recognizing the validity and possibilities inherent in organizing the world around different centers, and of acknowledging as fellow citizens other human beings with other faiths, we may somehow avoid some of the more pernicious consequences of the extreme positions we see so often propounded in popular venues today.

There is no way to demonstrate conclusively that a calculative approach to the world, even supported by a philosophical/scientific position which argues for the "truth" of science, is ultimately the closest to reality or the most fruitful for the future of human endeavors. In this respect, scientists should not be able to "pull rank" when the questions under discussion move outside the realm of scientific knowledge in the strict sense. If it is a question about moral values or the future flourishing of human communities, the scientist, no matter how eminent, cannot claim any special authority. She should, of course, enter the debate–but as a citizen along with other citizens. Certainly we can point to many salutary effects, both practical and intellectual, of the scientific/technological outlook on the world. Yet with equal ease, we can also enumerate many detrimental effects. This recognition neither issues a special invitation nor removes the scientist from the guest list of the public sphere. Surely this is also the case with people whose lives are lived with a religious motivation or a belief in art or beauty, rather than scientific rationality, as a motivating principle.

From this point of view, it is necessary that scientists and others with strong fundamentalist views of any stripe make a distinction between their role as scientists/specialists and their role as citizens. *Qua* scientist, one may try to analyze and understand the world, including the human beings in it, from the perspective of sociobiology. However, *qua* citizen, one has no right to assert such a view as an absolute basis for making responsible decisions in the public sphere. On the other hand, *qua* religious believer, one may also try to analyze and understand the world, including the human beings in it, from the perspective of ones faith. Again, however, *qua* citizen, one has no right to assert such a view as an absolute basis for making responsible decisions in the public sphere. Fellow citizens with

other, and often equally respectable views, have every right
to protest such a usurpation of democratic discourse.[21]

In the limited realm of scientific research, it may make
sense to choose to believe in an idealized scientific truth.
This ideal both orients and gives meaning to the whole
scientific enterprise. However, do we need to give up either the practice of science or a belief in its importance
and power to choose also a position that affirms a belief
in, for example, human freedom? Can the belief in freedom be embraced for its life-affirming qualities, for its faith
in the ability of human beings to work for a different and
better world, even if this belief is acknowledged as contradictory to the position adopted when pursuing scientific
research? Kant argues that we need to act *as if* we were free
because otherwise we consign ourselves to a world where
only chance and natural causality operate. This is a world
in which there can be neither responsibility for our actions
nor any hope of humanly motivated change. In such a
world in the absence of human responsibility, we would
have only a past but no history in Arendt's sense of the
word. Thus, even scientists may need to reject a mechanistic world view for moral questions, or, at least they should
recognize that there may be good, *rational* reasons why
others reject an essentially calculative orientation toward
the world.

Scientism attempts to use the force of fact effectively to
negate the fecundity of rich public dialogue. That is, scientism
says "here are the facts," and these facts are meant to compel
us to assent, bypassing the need for dialogue and careful
persuasion. Granted, at a scientific conference presenting data
this way and focusing minimally on the rhetorical aspects of
the presentation of facts, findings, and data may indeed compel others to productive and important ends. But in the public
sphere where questions of human futures are debated, facts
although necessary, alone cannot compel assent. Furthermore,
expectations of the same rules and procedures in the public
sphere that one has in legitimate scientific enterprises is a key

component of scientism. A more subtle and difficult form of human being-together is required where questions of ethics and public policy are concerned.

In the coming years, many problems with a scientific component will require such a subtle and difficult discourse. Among them are genetic engineering (Frankenfoods), stem cell research, human cloning, global warming, deforestation, ozone layer depletion, space-based defense (Star Wars), energy efficiency, and many others on an ever-growing list. Cautionary voices, some from concerned scientists, are currently raised regarding all these areas, but they have little force because scientism and its rhetorical strategies and prestige are so overwhelmingly aligned with corporate and military interests instead of with the public good.

Leaving Us to Wonder

The most consequential and important meanings of scientific theories such as evolution begin precisely at that point where the facts of evolution end. However, there is no scientific way to connect this ending point of facts with the beginning point of an ethical understanding of meaning and conduct. This requires another discourse, a shift of orientation, and a different grounding for our thinking. Nowhere is this requirement more important than in the public sphere.

We do not offer our critique with any intention of advocating a halt to scientific research, nor as a call to teach Creationism in public schools (quite the opposite), nor to harken back in some nostalgic way for a return to some premodern fantasy when all was right with the world before the advent of modern science and technology (there being no such time). To the contrary, we undertake these exercises of reflecting about science and reflecting about other ways of thinking to make a space where we can make a way of life. By all means we should take up scientific data into our thinking about life, thinking about mathematics,

thinking about mechanism and materialism to their ultimate theoretical and experimental ends. Yet, do not forget
that to make them a part of a way of life—to choose freely
a way—requires a type of thinking that departs from calculative methods and starts again on fresh ground.

We turn again to philosophy's beginning in wonder
and to our belief in the necessity of meditative thinking.
Curiosity about the world may lead us to investigate subjects
like the structure and function of DNA. It may also lead to
theorizing about the role DNA may play in our social and
political life. Wonder, on the other hand, leads us to investigate ourselves and our experiential lives and to think about
how our actions are thereby affected. But, wonder is not
ultimately about answers, it is about eternal questions. Arendt
puts a fine point on all that we have advocated throughout
this work when she writes of what it means to return to
engage communicatively with other human beings after
having withdrawn into the wonder that is the beginning of
all philosophy:

> As soon as the speechless state of wonder trans-
> lates itself into words, it will not begin with state-
> ments but will formulate in unending variations
> what we call the ultimate questions—What is be-
> ing? Who is man? What meaning has life? What is
> death? etc.—all of which have in common that
> they cannot be answered scientifically. . . .[22]

Here statements are not the order of the day. One does not
return from a passionate consideration in the face of wonder with facts or definitive statements, rather such an endeavor engenders the humility of questioning. More than
this it shows us who we are, namely, beings who can ask and
indeed must ask questions that have no final answers. The
stakes for ignoring or attempting to prohibit such questions
would mean the end of *human* life:

> Were man ever to lose the faculty of asking unan-
> swerable questions, he would by the same token
> lose his faculty of asking answerable questions. He
> would cease to be a question-asking being, which
> would be the end, not only of philosophy, but of
> science as well.[23]

Thus, we see, despite the protests of scientism against the utility of meditative thinking and scientism's unending claims for intellectual superiority over all types of thinking, it remains that the very questions science cannot ask and consequently cannot answer are constitutive of the genuine structure of human being-in-the-world that makes science possible at all.

Hadot ends his study of philosophy as a way of life warning of dangers that await those of us who put into practice the strategies for thinking and living we advocate. The danger he considers most pressing "is to believe that one can do without philosophical reflection. The philosophical way of life must be justified in rational, motivated discourse, and as such discourse is inseparable from a way of life."[24] It is for this reason that we have turned here at the end of our study to the reclaiming of the public sphere as a place where such discourse could flourish and perspectives could be presented and judged by all those willing to participate. We do not, of course, undertake this task of philosophical reflection and discourse for private reasons. As Hadot so eloquently puts it: "despite the tenacious clichés which still clog philosophy manuals, we must never forget that ancient philosophical life was always intimately linked to the *care for others*, and that this demand is inherent in the philosophical way of life, especially when it is lived *in the contemporary world*."[25] Thus, we end with the beginning of a permanent task always before us.

Prior to as well as beyond all facts, thinking places us before wonder and the question of living well.

Notes

Chapter One

1. For a detailed analysis of the notion of origin and its relation to where thinking begins, see Calvin O. Schrag, "Origins of the Human Sciences," in *Philosophical Papers: Betwixt and Between* (Albany: State University of New York Press, 1994).

2. John Dewey, *Experience and Nature* (New York: Dover, 1929), p. 192.

3. Dewey, *Experience and Nature*, pp. 410–411.

4. John Horgan, "The New Social Darwinists," *Scientific American* 273 (October 1995): 174–81.

5. Daniel Dennett, *Darwin's Dangerous Idea* (New York: Touchstone Books, 1995), pp. 519–520.

6. E. O. Wilson, *Sociobiology* (Cambridge: Harvard University Press, 1975); James Lovelock, *The Ages of Gaia: A New Look at Life on Earth* (Oxford: Oxford University Press, 1979); and Humberto Maturana and Francisco Varela, *The Tree of Knowledge: The Biological Roots of Human Understanding* (Boston: Shambala, 1992).

7. Philip Kitcher, *Vaulting Ambition: Sociobiology and the Quest for Human Nature* (Cambridge: MIT Press, 1985).

8. Richard Dawkins, "God's Utility Function," in *River Out of Eden* (New York: Basic books, 1995), pp. 95–133.

9. Ibid. p. 98.

10. For an account and reproduction of many of the relevant exchanges surrounding this issue see *The Sokal Hoax: The Sham that Shook the Academy* (Lincoln: University of Nebraska Press, 2000).

11. Alan Sokal and Jean Bricmont, *Fashionable Nonsense: Postmodern Intellectuals' Abuse of Science* (New York: Picador Press, 1998), p. 188.

12. We cite here three definitions and discussions of scientism to be read along with ours to gain a sense of the outlook we have throughout this work.

For John Dupree scientism is: "an exaggerated and often distorted conception of what science can be expected to do or explain for us. One aspect of scientism is the idea that any question that can be answered at all can best be answered by science." John Dupré, *Human Nature and the Limits of Science* (Oxford: Clarendon Press, 2001), p. 2–3.

For Neil Postman scientism is understood as: "three interrelated ideas. . . . The first and indispensable idea is, as noted, that the methods of the natural sciences can be applied to the study of human behavior. . . . The second idea is, as also noted, that social science generates specific principles which can be used to organize society on a rational and humane basis. This implies that technical means–mostly 'invisible technologies' supervised by experts–can be designed to control human behavior and set it on the proper course. The third idea is that faith in science can serve as a comprehensive belief system that gives meaning to life, as well as a sense of well-being, morality, and even immortality." Neil Postman, *Technopoly: The Surrender of Culture to Technology* (New York: Vintage Books, 1993), p. 147.

For Simon Critchley: "The critique of scientism resides in the belief that the model of the natural sciences cannot and, moreover, should not provide a model for philosophical method, and that the natural sciences do not provide human beings with their primary and most significant access to the world. One finds this belief expressed by a whole range of Continental thinkers, such as Bergson, Husserl, Heidegger and the philosophers associated with the Frankfurt School from the 1930s onwards."

Later he writes: "The critique of scientism within phenomenology does not seek to refute or negate the results of scientific research in the name of some mystical apprehension of the unity of man and nature, or whatever. Rather it simply insists that science does not provide the most primary or most significant access to a sense of ourselves and the world." Simon Critchley, *Continental Philosophy: A Very Short Introduction* (Oxford: Oxford University Press, 2001), pp. 111–112, and 115.

13. John Lyne, "Bio-Rhetorics: Moralizing the Life Sciences," in *The Rhetorical Turn: Invention and Persuasion in the Conduct of Inquiry*, ed. Herbert Simons (Chicago: University of Chicago Press, 1990), p. 39.

Chapter Two

1. Lucretius, *On the Nature of the Universe*, trans. Ronald Melville (Oxford: Oxford University Press, 1997), p. 54.

2. Galileo Galilei, *Dialogue Concerning the Two Chief World Systems*, trans. Stillman Drake (Berkeley: University of California Press, 1967).

3. Though Galileo's writing does not challenge belief in the Christian God or the authority of the Church on religious matters, the Roman

Catholic Church in his time officially supported Aristotelian physics—foremost here perhaps their sanctioning the belief that heavier bodies fall to earth faster, the Earth is the center of the universe, and that matter is not composed of atoms. Clearly the Church saw their divine teachings threatened by such un-Aristotelian doctrines as Copernicanism and atomism. People were even condemned as heretics and executed for advocating an atomic view of the composition of the material world. Here again, theories about the material world were seen as intimately connected with issues far beyond their apparent sphere. This legacy is still apparent today, though of course with its own contemporary particularities and details. Galileo's own fate is recorded in history and in the popular imagination. Such challenges and their influence had implications for others writing and thinking in this vein, including thinkers whose work has also had lasting consequences for contemporary scientific thought.

4. René Descartes, *Meditations on First Philosophy and Meditations,* trans. (New York Modern Library, 1960), p. 114.

5. Immanuel Kant, *Critique of Pure Reason,* trans. Norman Kemp Smith (New York: St. Martin's Press, 1963).

6. Immanuel Kant, *Critique of Practical Reason,* trans. Lewis White Beck (Indianapolis: Bobbs-Merrill, 1956).

7. See the Institute for Creation Research: http://www.icr.org/

8. Phillip Johnson. *Darwin on Trial* (Washington D.C.: InterVarsity Press, 1993).

9. For example see, Mary Midgeley, *Evolution as a Religion* (New York: Methuen, 1985), and *Science as Salvation* (New York: Routledge and Kegan Paul, 1992).

10. Robert Wright, "Our Cheating Hearts," *Time,* 15 Aug. 1994.

11. Robert Wright, "Feminists, Meet Mr. Darwin," *The New Republic,* 28 Nov. 1994, 34-46, and "The Biology of Violence," *The New Yorker,* 13 March 1995.

12. Some of the too numerous to exhaust list might include: E. O. Wilson, *On Human Nature* (1975); Richard Dawkins, *The Selfish Gene* (1990), *The Blind Watchmaker* (1987), and *River Out of Eden* (1995); Richard Alexander, *Darwinism and Human Affairs* (1979); Robert Axelrod, *The Evolution of Cooperation* (1985); and Daniel Dennett, *Darwin's Dangerous Idea* (1995). A slightly different tack, which includes the environmental factor of birth order along with the standard Darwinian analysis is taken by Frank Sulloway in *Born to Rebel* (1996). No doubt this list of books grows yearly as examples in magazines grow monthly and in newspapers such examples seem to appear nearly every day.

13. Wright, "Our Cheating Hearts," p. 46.

14. Ibid., p. 49.

15. Ibid., p. 50.

16. Ibid., p. 50.

17. Robert Wright, *The Moral Animal* (New York: Pantheon, 1994), pp. 365-366.

18. Wright, *The Moral Animal,* pp. 365–366.

19. Susan Blackmore, *The Meme Machine* (Oxford: Oxford University Press, 1999), p. 237.

20. See Howard L. Kaye, *The Social Meaning of Modern Biology* (New Haven: Yale University Press, 1986).

21. Simpson, G. G., "The Biological Nature of Man," *Science* 152, no. 3721 (1966) 472–478. This same text is quoted approvingly in Randy Thornhill and Craig Palmer's *A Natural History of Rape: Biological Bases of Sexual Coercion* (Boston: MIT Press, 2000), a book and argument with which we deal at length in chapter 6.

22. Richard Dawkins, *The Selfish Gene* (Oxford: Oxford University Press, 1989) p. 3.

23. Ibid., p. 267.

Chapter Three

1. Charles Darwin, *The Origin of Species* (New York: Penguin, 1968).

2. Darwin, *Origin,* p. 137.

3. Charles Darwin, *Descent of Man* (Princeton, NJ: Princeton University Press, 1981).

4. Friederich Nietzsche, *Beyond Good and Evil,* trans. Walter Kaufmann (New York: Vintage Books, 1966); *The Birth of Tragedy,* trans. Walter Kaufmann (New York: Vintage Books, 1967); *Genealogy of Morals,* trans. Walter Kaufmann (New York: Vintage Press, 1969); and *Thus Spoke Zarathustra,* trans. R. J. Hollingdale (Middlesex: Penguin Books Ltd, 1961).

5. Darwin, *Descent,* p. 51, 32, and 216.

6. Nietzsche, *Genealogy of Morals,* p. 11.

7. Darwin, *Descent,* p. 112.

8. Ibid., p. 98.

9. Darwin, *Descent,* p. xx.

10. Ibid., pp. 403–404.

11. Ibid., *Descent,* p. 104.

12. Nietzsche, *Genealogy,* p. 85.

13. Ibid., p. 117.

14. Ibid., pp. 48–52.

15. Ibid., pp. 146–147.

16. Ibid., pp. 155–156.

17. Ibid., p. 151.

18. Ibid., p. 151.

19. Ibid., pp. 148–149.

20. Ibid., p. 153.
21. Ibid., p. 20.
22. Ibid., p. 24.
23. Ibid., p. 3.
24. Ibid., p. 15.

Chapter Four

1. Martin Heidegger, *Zollikon Seminars: Protocols-Conversations-Letters*, ed. Medard Boss, trans. Franz Mayr and Richard Askay (Evanston, IL: Northwestern University Press, 2001), p. xv.

2. Heidegger, *Zollikon Seminars*, p. xviii.

3. Ibid., p. xviii. (Italics added)

4. As with all of the philosophers with whom we have dealt and as the above demonstrates, Heidegger's work is detailed and complicated and is contained in many volumes. As with the other thinkers with whom we are in dialogue here, we seek only to highlight those aspects of his thought that we feel most directly bear on our concerns.

5. Heidegger, "Memorial Address," in *Discourse on Thinking*, trans. John M. Anderson and E. Hans Freund (New York: Harper and Row, 1966), p. 46. (Italics added)

6. Heidegger, "Memorial Address," p. 47.

7. Martin Heidegger, "Science and Reflection," in *The Question Concerning Technology*, trans. William Lovitt (New York: Harper and Row, 1977).

8. Heidegger, "Science and Reflection," in *Question Concerning Technology*, p. 173.

9. Heidegger, "Science and Reflection," p. 174.

10. Heidegger, " Age of the World Picture," in *Question Concerning Technology*, p. 122.

11. Heidegger, "Science and Reflection," p. 168.

12. Ibid., p. 167–168. (Italics added)

13. Heidegger, "Memorial Address," p. 52–53.

14. Heidegger, *Zollikon Seminars*, p. 27.

15. Martin Heidegger, *What Is Called Thinking?*, trans. Fred D. Wieck and J. Glenn Gray (New York: Harper and Row, 1968), p. 42.

16. Maurice Merleau-Ponty, *The Phenomenology of Perception*, trans, Colin Smith (Atlantic Highlands, New Jersey: Humanities Press, 1962).

17. See Drew Leder, *The Absent Body* (Evanston, IL: Northwestern University Press, 1990).

18. Merleau-Ponty, *Phenomenology of Perception*, pp. 125 ff.

19. Robert Melzack, "Phantom Limbs," *Scientific American* 256: 120–126.

20. Friederich Nietzsche, *Beyond Good and Evil*, trans. Walter Kaufmann (New York: Vintage, 1966), p. 16.

Chapter Five

1. See, for example, Mary Midgley, *Evolution as a Religion* (London: Routledge and Kegan Paul, 1985); and Robert Wesson, *Beyond Natural Selection*, (Boston: MIT Press, 1993).

2. "Facial Shape and Judgments of Female Attractiveness," *Nature* (1994).

3. www.uni-saarland.de/fak5/ronald/online

4. Audrey C. Chen, Craig German, and Dahlia W. Zaidel, "Brain Asymmetry and Facial Attractiveness: Facial Beauty is not Simply in the Eye of the Beholder," *Neuropsychologia* 35, no. 4, (1997): 471–476.

5. Randy Thornhill and Nancy W. Thornhill, "Rape: An Evolutionary Analysis" 1983, *Ethology and Sociobiology* 4: 137–173.

6. Thornhill and Thornhill, "Rape," p. 137.

7. Ibid., p. 167.

8. Ibid., p. 167

9. Furthermore, it is not surprising that the Thornhills reach the conclusion that they do when we consider their definition of rape (p. 141), which hinges not on whether the woman consents, but on whether the woman can use the "copulation" to her own advantage. Thus, if a boss has nonconsensual sex with his secretary, but she gets a raise and job security as a result, thus using the copulation "in her own reproductive advantage," it is not considered a rape. Notice that a definition of this sort essentially guarantees the conclusion that evolutionary biology is the only discipline which can give an account of the phenomenon of rape. Certainly, the legal, sociobiological, and psychoanalytic definitions are very different from this.

10. Randy Thornhill and Craig Palmer, *A Natural History of Rape: Biological Bases of Sexual Coercion* (Boston: MIT Press, 2000).

11. This book has been criticized by scientists, feminists, public health officials, and many others as poor science, a narrow view even from an evolutionary standpoint, and all together irresponsible in its scholarship and claims. see Susan Brownmiller, "Rape on the Brain," *Feminista* 3, no. 9 (2000); Barbara Ehrenreich, "How 'Natural' is Rape?" *Time*, 31 Jan 2000, vol. 155, no. 4.; and Jerry Coyne and Andrew Berry, "A Theory that Rape has its Origin in Evolutionary Biology is Seriously Flawed," *Nature* 404 (2000).

12. Peggy Reeves Sanday, "The Sociocultural Context of Rape: A Cross-Cultural Study," *Journal of Social Issues* 37, (1981): 5–27.

13. Thornhill and Palmer, *Rape*, p. 183.

Chapter Six

1. Pierre Hadot, *What is Ancient Philosophy?* trans. Michael Chase (Cambridge: Harvard University Press, 2002), p. 275.

2. See Hannah Arendt, "The Crisis in Culture," in *Between Past and Future* (New York: Penguin, 1954), pp. 225–226. She writes: "We can rise above specialization and philistinism of all sorts to the extent that we learn how to exercise our taste freely. Then we shall know how to reply to those *who so frequently tell us that Plato or some other great author of the past has been superseded*; we shall be able to understand that even if all criticism of Plato is right, Plato may still be better company than his critics. At any rate, we may remember what the Romans—the first people that took culture seriously the way we do—thought a cultivated person ought to be: one who knows how to choose his company among men, among things, among thoughts, in the present as well as in the past." (Italics added)

3. Hannah Arendt, "What is Freedom?" in *Between Past and Future* (New York: Penguin, 1954), p. 170–171.

4. Arendt, "What is Freedom?" p. 171.

5. We can also, as Nietzsche points out, invent theoretical worlds and then do our best to live in these imaginary worlds of our own creation. For him, Christianity and science both posit such theoretical worlds as their ideal. If we act in accordance with these beliefs, we thereby affect the future and alter, in effect, the history of the world. At least, we change ourselves so as to experience and conceive the world differently. To see the future wholly through the lens of science and technology is for us the danger of technocratic authoritarianism. This is not to overlook that a thin but dogmatic religious view held to the exclusion of all else is also a danger of immense consequence. Were we to continue to train ourselves to limit the ways the world can appear to us, then we would continue to act in the world in accordance with this limited vision and eventually the richness of nature and our own being would be closed off to us.

6. We have occasionally alluded to the problem of language, a problem that from Plato to the present philosophers see as vital to any understanding of value. When addressing issues of public deliberation it becomes an essential concern. In the *Genealogy of Morals* discussed in chapter three, Nietzsche showed how meanings and perhaps even the things themselves could be changed dramatically through the use of language. We can call an old thing like "weakness" by a new name like "humility" and elaborate a whole new religion and cultural traditions based on this new use of language. Based in this type of insight, contemporary thinkers take scientists to task for their naiveté concerning the uses of language. This becomes especially important when scientists write for the general public. Many critics have addressed this issue. One powerful example is the work of Johyn Lyne and Henry F. Howe, "The Rhetoric

of Expertise: E. O. Wilson and Sociobiology" *Quarterly Journal of Speech* 1990 (76) pp. 134-151, who ask how a public with no formal scientific education is going to understand the texts published for their benefit except through using the resources that they have? Though scientists will often pay lip service to the strict scientific meaning of terms like "selfish" or "gene," their constant use of these terms in wide, general contexts shows that they themselves neither understand nor use such terms in their "strict" meanings. One of the most important and fascinating aspects of language is our inability to control completely and without remainder the meaning of words. They are almost always larger than our intentions for them, even when we consciously try to restrict them. This is the basis for the richness of poetry, psychoanalysis, humor, and much of everyday life.

7. Immanuel Kant, "What is Orientation in Thinking?" in *Political Writings*, trans. H. B. Nisbet (Cambridge: Cambridge University Press, 1970), p. 247.

8. Kant, "What is Orientation in Thinking?", p. 247.

9. Allow us another excursus regarding Nietzsche, one that leads to Daniel Dennett. Nietzsche has argued that an irrational faith in an idealized metaphysical truth, one that stems from our deepest needs to find meaning in the world, underlies both the belief in the Christian God and the belief in modern science. In both these forms, asceticism has stood as the road to power over the minds and souls of the general population. This asceticism, for Nietzsche, is an inchoate desire for the annihilation of our "evil" animal selves which shows itself in modern evolutionary theory as a belief that the real meaning of life, even human life, is the mindless replication of the chemical substance DNA, just as humans, in the ascetic Christian view, are seen as nothing before the awesome greatness of God.

Following Nietzsche's lead, we can see very many additional parallels between these two positions. The ascetic priest of Christianity, through his special knowledge, is able to tell his flock the truth about their sinful nature and to offer them salvation if and only if they believe absolutely in God and follow His dictates, otherwise eternal torment in hell awaits them.

In a strikingly parallel manner, scientists assure us that they know the deep hidden truths of our sinful nature—it is in our (selfish) genes. They also promise us salvation for humanity and indeed the entire world if and only if we abandon our false gods and believe only in the ways of science. Failure to follow this path, they assure us, will result in our destruction through health and environmental evils, and our inability to control our own evil and destructive natures. Such control, we are told, can only come about when we admit the truth of our selfish gene-driven natures.

We may at first be puzzled as to why our attention is so continually drawn to the contemplation of DNA. It may be true that, for instance, art evolved as a means of sexual display (see the October 1995 issue of

Scientific American), but why should not the beauty and diversity of artistic productions capture our attention in their own right? Why should we even care about its alleged ultimate genetic origin? In these proudly proclaimed secular and scientific texts, we can almost hear Augustine shouting from the shadows, telling us that it is evil to worship created things rather than the Creator, that it is evil to apply our meanings to things rather than the ultimate meaning of our Creator—difficult as it may be to understand and contrary as it may be to our desires.

We have seen that it is possible to construct imaginative mechanistic accounts that are claimed to be adequate ultimate explanations of all phenomena. However, when these accounts rely on concepts with little or no empirical support and rest on an unacknowledged metaphysical foundation—should we embrace them? Why should we accept, for example, Daniel Dennett's claims that the *St. Matthew's Passion* is a random "meme" in "design space" discovered by J. S. Bach and which desires to perpetuate itself "for its own sake" by "infecting" human brains, rather than seeing it as ultimately a sublime and purposeful expression of religious feeling and understanding? Just because it is possible to interpret it according to memes, this hardly seems a strong enough recommendation to override our own experience and put in its place such a banal interpretation.

The evolutionary view of human nature has a problem. Unless they draw strong but questionable conclusions, for example, that everything is really selfishness, from what are claimed to be scientific foundations, then they cannot claim that biology should play an essential role in our evaluations of ethical or cultural questions. Let's say we believe in "memes" that are able to override the dictates of selfish genes. Now we do not have to invoke biology for any troubling or seemingly contradictory instance of behavior. Once we are free from this constraint, we can invoke genes or memes as convenience dictates and no conclusions can follow with any compelling power from the biological facts. At this point we see that desperate appeals to concepts such as memes are no more than attempts at protecting the faith, that is to say, we could have no possible need for such an abstract concept unless we were previously committed to an exclusive belief in materialist forces. See, Daniel Dennett, *Darwin's Dangerous Idea* (New York: Touchstone Press, 1995).

10. Vaclav Havel, "Politics and Conscience," in *Living in Truth* ed. Jan Vladislav (New York: Faber and Faber, 1984), p. 146.

11. Havel, "Politics and Conscience," p. 146.

12. Havel, "Politics and Conscience," p. 149. (Italics added)

13. Vaclav Havel "Politics, Morality, Civility," in *Summer Meditations* (New York: Vintage, 1993).

14. Hannah Arendt, "On Humanity in Dark Times," in *Men in Dark Times* (New York: Harcourt Brace and Company, 1983).

15. Joseph L. Graves, Jr., *The Emperor's New Clothes: Biological Theories of Race at the Millennium* (New Brunswick: Rutgers University Press, 2001).

16. Graves, *The Emperor's New Clothes*, p. 197.

17. Ibid., p. 2.

18. See, for example, Hannah Arendt, "Thinking and Moral Considerations," *Social Research* 38, no. 3 (1971): 7–37.

19. This position is strikingly and brilliantly laid out by Jacques Derrida. It bears quoting at length as it helps to get at the heart of our understanding of the situation facing us:

> Far from opposing undecidability to decision, I would argue that there would be no decision, in the strong sense of the word, in ethics, in politics, no decision, and thus no responsibility, without the experience of some undecidability. If you don't experience some undecidability, then the decision would simply be the application of a programme, the consequence of a premiss or a matrix. So a decision has to go through some impossibility in order for it to be a decision. If we knew what to do, if I knew in terms of knowledge what I have to do before the decision, then the decision would not be a decision. It would simply be the application of a rule, the consequence of a premiss, and there would be no problem, there would be no decision. Ethics and politics, therefore, start with undecidability. I am in front of a problem and I know that the two determined solutions are as justifiable as one another. From that point, I have to take responsibility which is heterogeneous to knowledge. If the decision is simply the final moment of a knowing process, it is not a decision. So the decision first of all has to go through a terrible process of undecidability, otherwise it would not be a decision, and it has to be heterogeneous to the space of knowledge. If there is a decision it has to go through undecidability and make a leap beyond the field of theoretical knowledge ... Not knowing what to do does not mean that we have to rely on ignorance and to give up knowledge and consciousness. A decision, of course, must be prepared as far as possible by knowledge, by information, by infinite analysis. At some point, however, for a decision to be made you have to go beyond knowledge, to do something that you don't know, something which does not belong to, or is beyond, the sphere of knowledge. That is why the distinction between good and evil doesn't depend on knowledge; that is why we should not know, in terms of knowledge, what is the distinction between good and evil. To have to make such a distinction, which depends precisely on responsibility, is, I confess, both a terrible and tragic

situation in which to find oneself. Without this terrible experience, however, there would be no decision, there would simply be a serene application of a programme of knowledge and then we could delegate decisions to scientists and theoreticians.

"Hospitality, Justice, and Responsibility: A Dialogue with Jacques Derrida" in *Questioning Ethics: Contemporary Debates in Philosophy* eds. Richard Kearney & Mark Dooley (New York: Routledge, 1999), 66–67.

20. There are many other arguments that should make us hesitate for some deep reflection before enthusiastically embracing the evolutionary world view. In addition to Havel's and Arendt's claims here, Barry Schwartz shows how they serve as the foundation for exploitive economic practices and social relations. This should become a concern for us irrespective of the "truth" of such scientific claims. If we embrace such views wholeheartedly, such that we acknowledge no other possible independent lines of argument, it is not clear that we can ever refute the first principles from which the actions of such regimes flow. See, Barry Schwartz, *The Battle for Human Nature* (New York: W. W. Norton, 1986).

21. See Richard Rorty, "Religion as a Conversation Stopper," in *Philosophy and Social Hope* (London: Penguin Books Ltd, 1999), and his follow-up piece, "Religion in the Public Sphere: A Reconsideration," *Journal of Religious Ethics* 31–1: 141–149.

22. Hannah Arendt, "Philosophy and Politics," *Social Research* 57, no. 1 (Spring 1990): 98–99.

23. Ibid., p. 99.

24. Hadot, *What is Ancient Philosophy?* p. 280.

25. Ibid., p. 280. (Italics added)

Bibliography

Alexander, Richard. *Darwinism and Human Affairs*. Seattle: University of Washington Press, 1979.

Arendt, Hannah. "On Humanity in Dark Times" in *Men in Dark Times*. New York: Harcourt, Brace & World, 1968.

———. *Between Past and Future*. New York: Penguin, 1954.

———. "Thinking and Moral Considerations: A Lecture." *Social Research* 38, no. 3 (1971): 417–437.

———. *Eichmann in Jerusalem*. New York: Penguin, 1964.

———. *Origins of Totalitarianism*. New York: Harcourt, Brace & World, 1969. especially pp. 441–457.

Ashmore, Malcolm, Derek Edwards, and Jonathan Potter. "The Bottom Line: The Rhetoric of Reality Demonstrations." *Configurations* 1 (1994): 1–14.

Axelrod, Robert. *The Evolution of Cooperation*. New York: Basic Books, 1985.

Blackmore, Susan. *The Meme Machine*. Oxford: Oxford University Press, 1999.

Bugbee, Henry G. *The Inward Morning: A Philosophical Exploration in Journal Form*. Athens: University of Georgia Press, 1999.

Bergson, Henri. *Creative Evolution*. Translated by Arthur Mitchell. Lanham, MD: University Press of America, 1983.

Brownmiller, Susan. "Rape on the Brain." *Feminista*. 3, no. 9 (2000).

Chen, Audrey C., Craig German and Dahlia W. Zaidel. "Brain Assymetry and Facial Attractiveness: Facial Beauty is not Simply in the Eye of the Beholder." *Neuropsychologia*. 35, no. 4 (1997): 471–476.

Coyne, Jerry and Andrew Berry. "A Theory that Rape has its Origin in Evolutionary Biology is Seriously Flawed." *Nature.* 404 (2000).

Critchley, Simon. *Continental Philosophy: A Very Short Introduction.* Oxford: Oxford University Press, 2001.

Darwin, Charles. *Descent of Man and Selection in Relation to Sex.* Princeton: Princeton University Press, 1981.

———. *The Origin of Species.* New York: Penguin, 1968.

Dawkins, Richard. "God's Utility Function." *Scientific American* 38, no. 273 (November 1995): 80–85.

———. *The Blind Watchmaker.* New York: Norton, 1987.

———. *The Selfish Gene.* New York: Oxford, 1990.

———. *River out of Eden.* New York: Basic Books, 1995.

Dennett, Daniel. *Darwin's Dangerous Idea.* New York: Simon & Schuster, 1995.

Descartes, Rene. *Discourse on Method and Meditations.* Translated by Laurence J. LaFleur. New Jersey: MacMillian, 1960.

Dewey, John. *Experience and Nature.* New York: Dover, 1929.

Dupré, John. *Human Nature and the Limits of Science.* Oxford: Clarendon Press, 2001.

Ehrenreich, Barbara. "How 'Natural' is Rape?" *Time.* 155, no. 4 (January 2000).

Foucault, Michel. *Ethics: Subjectivity and Truth.* Translated by Robert Hurley and others. New York: The New Press, 1994.

Galileo. *Dialogue Concerning the Two Chief World Systems.* Translated by Stillman Drake. Berkeley: University of California Press, 1967.

———. *Two New Sciences: Including Centers of Gravity & Force of Percussion.* Translated by Stillman Drake. London: University of Wisconsin Press, 1974.

Gould, Stephen J. *The Mismeasure of Man.* New York: Norton, 1981.

Graves, Joseph L., Jr. *The Emperor's New Clothes: Biological Theories of Race at the Millenium.* New Brunswick: Rutgers University Press, 2001.

Gross, Paul R., and Norman Levitt. *Higher Superstition: The Academic Left and its Quarrels with Science.* Baltimore: Johns Hopkins University Press, 1994.

Havel, Vaclav. *Living in Truth.* Edited by Jan Vladislav. New York: Faber and Faber, 1984.

Heidegger, Martin. *Basic Writings.* Edited by David Farrell Krell. San Francisco: Harper & Row, 1977.

———. "Memorial Address" in *Discourse on Thinking.* Translated by John M. Anderson and E. Hauns Freund. New York: Harper & Row, 1966.

———. *Being and Time.* Translated by J. Macquarrie & E. Robinson. San Francisco: Harper, 1962.

———. *The Question Concerning Technology and Other Essays.* Translated by W. Lovitt. New York: Harper & Row, 1977.

———. *What is Called Thinking?* Translated by Fred D. Wieck and J. Glenn Gray. New York: Harper & Row, 1968.

———. *Zollikon Seminars: Protocols—Conversations—Letters.* Translated by Franz Mayr and Richard Askay. Evantson, IL: Northwestern University Press, 2001.

Hadot, Pierre. *Philosophy as a Way of Life: Spiritual Exercises from Socrates to Foucault.* Translated by Michael Chase. Oxford: Blackwell Publishers Ltd., 1995.

———. *What is Ancient Philosophy?* Translated by Michael Chase. Cambridge: The Belknap Press of Harvard University Press, 2002.

Heinsohn, Robert, and Craug Packer. "Complex Cooperative Strategies in Group: Territorial African Lions." *Science* 269, no. 5228 (September 1995).

Heller, Agnus. *Can modernity survive?* Berkeley: University of California Press, 1990.

Holton, Gerald. *Science and Anti-Science.* Cambridge: Harvard University Press, 1993.

Horgan, John. "The New Social Darwinists." 181 *Scientific American* (1995): 273.

Horkheimer, Max and Theodore Adorno. *Dialectic of Enlightenment.* Translated by John Cumming. New York: Continuum, 1986.

Hume, David. *Enquiries Concerning the Human Understanding and Concerning the Principles of Morals.* 1777. Reprint with introduction, comparative tables of contents, and analytical index by L. A. Selby-Bigge, Oxford: Clarendon Press, 1961.

Johnson, Phillip. *Darwin on Trial.* Washington D.C.: InterVarsity Press, 1993.

Kant, Immanuel. "An Answer to the Question: 'What is Enlightenment.'" In *Political Writings,* edited by H. S. Reiss, and translated by H. B. Nisbet. Cambridge: Cambridge University Press, 1970.

———. *Critique of Judgment.* Translated by James Creed Meredith. Oxford: Clarendon Press, 1952.

———. *Critique of Practical Reason.* Translated by Lewis White Beck. New York: Macmillan, 1956.

———. *Critique of Pure Reason.* Translated by Norman Kemp Smith. New York: St. Martin's, 1965.

Kaye, Robert. *The Social Meaning of Modern Biology.* New Haven: Yale University Press, 1986.

Kearney, Richard and Mark Dooley, eds. *Questioning Ethics: Contemporary Debates in Philosophy.* New York: Routledge, 1999.

Keller Fox, Evelyn. *Reflections on Gender and Science.* New Haven: Yale University Press, 1985.

Kitcher, Philip. *Vaulting Ambition: Sociobiology and the Quest for Human Nature.* Cambridge: MIT Press, 1985.

Koshland, Daniel. "A Milk-Free Zone." 264 *Science* (1994): 11.

Leder, Drew. *The Absent Body.* Chicago: University of Chicago Press, 1990.

Levins, Richard & Lewontin, R. C., *The Dialectical Biologist.* Cambridge: Harvard University Press, 1985.

Lewontin, R. C., Steven Rose, and Leon J. Kamin. *Not in Our Genes.* New York: Pantheon, 1984.

———. *The Sokal Hoax : The Sham that Shook the Academy.* Edited by the editors of *Lingua Franca.* Lincoln: University of Nebraska Press, 2000.

Lovelock, James. *The Ages of Gaia: A New Look at Life on Earth.* Oxford: Oxford University Press, 1979.

Lucretius. *On the Nature of the Universe.* Translated by Ronald Melville. Oxford: Oxford University Press, 1997.

Lyne, John and Henry F. Howe. "The Rhetoric of Expertise E. O. Wilson and Sociobiology." *Quarterly Journal of Speech.* 76 (1990): 134–151.

Maddox, John. "Defending science against anti-science." *Nature* 368 (March 1994): 185.

Majeski, Tom. "Myth of cooperative behavior within Pride is shattered by researchers: King of the jungle isn't so lion-hearted." *Arizona Republic* (1 Sept. 1995): A-1.

Markos, Anton. "The Ontogeny of Gaia: The Role of Microorganisms in Planetary Information Network." *Theoretical Biology* 176 (1995): 175–180.

Maturana, Humberto and Francisco Varela. *The Tree of Knowledge: The Biological Roots of Human Understanding.* Boston: Shambala, 1992.

Melzack, Robert. "Phantom Limbs." *Scientific American* (April 1992): 120–126.

Merleau-Ponty, Maurice. *The Phenomenology of Perception.* Translated by Colin Smith. New Jersey: Humanities Press, 1962.

Midgeley, Mary. *Evolution as a Religion.* New York: Methuen, 1985.

——— *Science as Salvation.* New York: Routledge, 1992.

Moore, John. *Science as a Way of Knowing.* Cambridge: Harvard University Press, 1993.

Nelkin, Dorothy & Lindee, M. Susan. *The DNA Mystique.* New York: W. H. Freeman, 1995.

———. *The Gene as a Cultural Icon.* New York: W. H. Freeman, 1995.

Nelkin, Dorothy. *Selling Science.* New York: W. H. Freeman, 1987.

Nietzsche, Friederich. *Beyond Good and Evil.* Translated by Walter Kaufmann. New York: Vintage Books, 1966.

———. *The Birth of Tragedy.* Translated by W. Leaufmann. New York: Vintage, 1967.

————. *On the Genealogy of Morals.* Translated by Walter Kaufmann New York: Random House, 1969.

————. *Thus Spoke Zarathustra.* Translated by R. J. Hollingdale. Middlesex: Penguin Books Ltd., 1961.

Noble, David. *World without Men.* Oxford: Oxford University Press, 1992.

Ochs, Elinor, Sally Jacoby and Patrick Gonzales. "Interpretive Journeys: How Physicists Talk and Travel Through Graphic Space." 2, no. 1 *Configurations* (1994): 151–171.

Perrett, D. I, K. A. May and S. Yoshikawa. "Facial Shape and Judgments of Female Attractiveness." 368 *Nature* (1994): 239–242.

Plato. *Collected Dialogues.* Edited by Edith Hamilton and Huntington Cains. New Jersey: Princeton University Press, 1961.

Pietro, Redondo. *Galileo: Heretic.* Translated by R. Rosenthal. Princeton: Princeton University Press, 1987.

Postman, Neil. *Technology: The Surrender of Culture to Technology.* New York: Vintage Books, 1993.

Reid, R. G. B. *Evolutionary Theory: The Unfinished Synthesis.* Ithaca: Cornell University Press, 1985.

Richard, Robert. *Darwin and the Emergence of Evolutionary Theories of Mind and Behavior.* Chicago: Chicago University Press, 1987.

Rorty, Richard. *Philosophy and Social Hope.* London: Penguin Books Ltd., 1999.

————. "Religion in the Public Sphere: A Reconsideration." *Journal of Religious Ethics.* 31-1, 141–149.

Rose, Steven. "The Rise of Neurogenetic Determinism." 373 *Nature* (1995): 380–382.

Ross, Andrew. "The Chicago Ganster Theory of Life." 35 *Social Text* (1993): 92–112.

Sanday, Peggy Reeves. "The Sociocultural Context of Rape: A Cross-Cultural Study." *Journal of Social Issues.* 37 (1981): 5–27.

Schrag, Calvin O. *Philosophical Papers: Betwixt and Between.* Albany: State University of New York Press, 1994.

Schwartz, Barry. *The Battle for Human Nature*. New York: W. W. Norton, 1986.

Shaw, G. B. *Back to Methuselah*. Vol. 2, *Bernard Shaw: Complete Plays with Prefaces*. New York: Dodd, Mead & Company, 1961.

Simons, Herbert, ed. *The Rhetorical Turn: Invention and Persuasion in the Conduct of Inquiry*. Chicago: University of Chicago Press, 1990.

Sokal, Alan, "Transgressing the Boundaries: Toward a Transformative Hermeneutics of Quantum Gravity." 46/47 *Social Text* (1996): 217–252.

Sokal, Alan and Jean Briemont. *Fashionable Nonsense: Postmodern Intellectuals' Abuse of Science*. New York: Picador Press, 1998.

Sulloway, Frank. *Born to Rebel: Birth Order, Family Dynamics, and Creative Lives*. New York:Pantheon, 1996.

Thornhill, Randy and Craig Palmer. *A Natural History of Rape: Biological Bases of Sexual Coercion*. Boston: MIT Press, 2000.

Thornhill, Randy & Nancy W. Thornhill. "Rape: An Evolutionary Analysis." *Ethology and Sociobiology* 4 (1983): 137–173.

Wellmer, Albrecht. *The Persistence of Modernity*. Translated by David Midgley. Cambridge: MIT Press, 1993.

Wesson, Robert. *Beyond Natural Selection*. Cambridge: MIT Press, 1993.

Wilson, E. O. *On Human Nature*. Cambridge: Harvard University Press, 1975.

———. *Sociobiology*. Cambridge: Harvard University Press, 1975.

Wolfe, Alan. "Up from Scientism." *The New Republic*, 23 November 1996, 29–35.

Wright, Robert. "Feminists, Meet Mr. Darwin." *The New Republic*, 28 November 1994, 34–46.

———. "Is the Unabomber Right?" *Time*, 28 August 1995, 50–57.

———. "The Biology of Violence." *The New Yorker*, 13 March 1995, 68–77.

———. *The Moral Animal*. New York: Pantheon, 1994.

———. "Our Cheating Hearts." *Time* 15 August 1994, 44–54.

Index